HVAC & R Hands on Troubleshooting

Jose C. Jimenez

Order this book online at www.trafford.com
or email orders@trafford.com

Most Trafford titles are also available at major online book retailers.

Print information available on the last page.

ISBN: 978-1-4907-6099-5 (sc)
ISBN: 978-1-4907-6101-5 (hc)
ISBN: 978-1-4907-6100-8 (e)

Library of Congress Control Number: 2015909126

Trafford rev. 08/18/2015

www.trafford.com

North America & international
toll-free: 1 888 232 4444 (USA & Canada)
fax: 812 355 4082

Contents

Introduction xxi
A Refrigeration Experience xxv
How My Interest in Refrigeration Started xxvii

01. Refrigerant Leaks (U.S. Coast Guard)
     EXPERIENCE 01-A 1
     EXPERIENCE 01-B 279

02. Beverage Cooler
     EXPERIENCE 02-A 3
     EXPERIENCE 02-B 283

03. Repair of an A/C Equipment
     EXPERIENCE 03-A 5
     EXPERIENCE 03-B 285

04. Defected Compressor
     EXPERIENCE 04-A 6
     EXPERIENCE 04-B 288

05. A/C Frozen Evaporator
     EXPERIENCE 05-A 7
     EXPERIENCE 05-B 291

06. Freezer Grounded
     EXPERIENCE 06-A 10
     EXPERIENCE 06-B 294

07. Merchant's Ship
     EXPERIENCE 07-A 12
     EXPERIENCE 07-B 296

08. Fish Products Walking Box Repair
     EXPERIENCE 08-A 13
     EXPERIENCE 08-B 298

09. Thermostatic Expansion Valve Pilot Type
     EXPERIENCE 09-A 15
     EXPERIENCE 09-B 300

10. Frost and Defrost Repeated
     EXPERIENCE 10-A 17
     EXPERIENCE 10-B 301

11. Tropicalization?
     EXPERIENCE 11-A...19
     EXPERIENCE 11-B... 303

12. What's Wrong (DR)
     EXPERIENCE 12-A 21
     EXPERIENCE 12-B 305

13. Stored Meat Spoiled
     EXPERIENCE 13-A 23
     EXPERIENCE 13-B 307

14. Refrigeration System Contamination
     EXPERIENCE 14-A 25
     EXPERIENCE 14-B 309

15. Ice Blocks Factory
     EXPERIENCE 15-A 27
     EXPERIENCE 15-B 311

16. Restaurant Lower Manhattan: The A/C system Is Not Cooling
     EXPERIENCE 16-A 29
     EXPERIENCE 16-B 314

17. Walking Box Storage Oranges
    EXPERIENCE 17-A...32
    EXPERIENCE 17-B..316

18. Chemical Reaction
    EXPERIENCE 18-A...34
    EXPERIENCE 18-B..322

19. Columbus Avenue Bank: Repair That Wasn't Supposed to Be
    EXPERIENCE 19-A...36
    EXPERIENCE 19-B..323

20. Bank: Continuation of a Job?
    EXPERIENCE 20-A...37
    EXPERIENCE 20-B..324

21. CBS: A Frozen Evaporator
    EXPERIENCE 21-A...39
    EXPERIENCE 21-B..326

22. Rooftop: Cooling Tower's Fan Motor Mounting
    EXPERIENCE 22-A...41
    EXPERIENCE 22-B..328

23. Rooftop: Flooding Roof
    EXPERIENCE 23-A...42
    EXPERIENCE 23-B..329

24. Computer's Room: Change a Compressor
    EXPERIENCE 24-A...43
    EXPERIENCE 24-B..330

25. Museum Restaurant: Helping to Find a Refrigerant Leak
    EXPERIENCE 25-A...45
    EXPERIENCE 25-B..334

26. Fashion District: Flooding Cooling Tower
    EXPERIENCE 26-A 47
    EXPERIENCE 26-B 336

27. Madison Avenue: No A/C Air Supply After Regular Working Hours
    EXPERIENCE 27-A 48
    EXPERIENCE 27-B 337

28. A/C Equipment: Summer Preparation
    EXPERIENCE 28-A 50
    EXPERIENCE 28-B 338

29. O/L Wrong Electrical Installation
    EXPERIENCE 29-A 52
    EXPERIENCE 29-B 339

30. Carnegie Delicatessen: Compressor Missing Identification
    EXPERIENCE 30-A 54
    EXPERIENCE 30-B 340

31. Water Pump Impeller Stuck
    EXPERIENCE 31-A 55
    EXPERIENCE 31-B 341

32. Sharing Working Experience
    EXPERIENCE 32-A 56
    EXPERIENCE 32-B 342

33. A/C Computer's Room Frozen Evaporator
    EXPERIENCE 33-A 58
    EXPERIENCE 33-B 343

34. Care Building A/C Systems Are Not Cooling
    EXPERIENCE 34-A 61
    EXPERIENCE 34-B 348

35. Defrost Room Too Cold
    EXPERIENCE 35-A 63
    EXPERIENCE 35-B 351

36. Village: A/C System Is Not Cooling
    EXPERIENCE 36-A... 65
    EXPERIENCE 36-B 353

37. Italian Restaurant: A/C Maintenance Service
    EXPERIENCE 37-A 67
    EXPERIENCE 37-B... 356

38. Times Square Pharmacy: Preparing the A/C System for Summer
    EXPERIENCE 38-A... 69
    EXPERIENCE 38-B 357

39. Harvard Club: Refrigerant Recovery
    EXPERIENCE 39-A 71
    EXPERIENCE 39-B 358

40. Fish Market: The Crusher Ice/Maker Is Not Working
    EXPERIENCE 40-A... 73
    EXPERIENCE 40-B 361

41. Italian Restaurant: A/C System Is Not Cooling
    EXPERIENCE 41-A 75
    EXPERIENCE 41-B... 363

42. Fashion Warehouse: Flooded Cooling Tower
    EXPERIENCE 42-A... 76
    EXPERIENCE 42-B 365

43. Italian Restaurant: Condensing Unit Replacement
    EXPERIENCE 43-A... 78
    EXPERIENCE 43-B 366

44. Times Square Pharmacy: A/C System Is Not Cooling
    EXPERIENCE 44-A.. 80
    EXPERIENCE 44-B.. 368

45. Friars Club: Compressor's Change
    EXPERIENCE 45-A.. 82
    EXPERIENCE 45-B.. 369

46. Bank 16 St.: Unnecessary Change of Compressor
    EXPERIENCE 46-A.. 84
    EXPERIENCE 46-B.. 370

47. Italian Restaurant: A/C Not Cooling
    EXPERIENCE 47-A.. 87
    EXPERIENCE 47-B..374

48. Long Island: Service
    EXPERIENCE 48-A.. 89
    EXPERIENCE 48-B.. 377

49. "Student's Consult" California
    EXPERIENCE 49-A.. 91

50. Chuck-Full-of-Nuts: Service Call
    EXPERIENCE 50-A.. 93
    EXPERIENCE 50-B.. 378

51. La Grenouille: Heating Service Call
    EXPERIENCE 51-A.. 95
    EXPERIENCE 51-B.. 380

52. Empire Restaurant: Thermostat Misalignment
    EXPERIENCE 52-A.. 97
    EXPERIENCE 52-B.. 381

53. Student: A/C Transaction
EXPERIENCE 53-A 98
EXPERIENCE 53-B 382

54. Computer's Room: Refrigerant Leak
EXPERIENCE 54-A... 100
EXPERIENCE 54-B 384

55. Mistake: Refrigerant Charge
EXPERIENCE 55-A 102
EXPERIENCE 55-B 386

56. Computer Room: A/C System Does Not Start
EXPERIENCE 56-A... 104
EXPERIENCE 56-B 388

57. Computer's Room: A/C System Is Not Cooling
EXPERIENCE 57-A 106
EXPERIENCE 57-B... 390

58. WPIX: A/C Not Cooling
EXPERIENCE 58-A... 108
EXPERIENCE 58-B 392

59. WPIX: A/C Repair
EXPERIENCE 59-A 110
EXPERIENCE 59-B 394

60. Crank Case Broken Sight Glass
EXPERIENCE 60-A...112
EXPERIENCE 60-B 395

61. Dishonest Behavior
EXPERIENCE 61-A...114
EXPERIENCE 61-B... 398

62. A/C System Not Cooling (Pharmacy)
   EXPERIENCE 62-A 116
   EXPERIENCE 62-B 399

63. Wiring Electrical Installation
   EXPERIENCE 63-A 119
   EXPERIENCE 63-B 402

64. Automobile A/C Repairs
   EXPERIENCE 64-A 121
   EXPERIENCE 64-B 404

65. Adding Three Water-Cooled Condensers
   EXPERIENCE 65-A 123
   EXPERIENCE 65-B 406

66. Long Time to Start
   EXPERIENCE 66-A 125
   EXPERIENCE 66-B 409

67. Exhaust Fan
   EXPERIENCE 67-A 128
   EXPERIENCE 67-B 412

68. Settlement Tank
   EXPERIENCE 68-A 129
   EXPERIENCE 68-B 413

69. Installation of a Walk-In Freezer
   EXPERIENCE 69-A 130
   EXPERIENCE 69-B 414

70. Supermarket Frosted Suction Line
   EXPERIENCE 70-A 131
   EXPERIENCE 70-B 417

71. High-Discharge Pressure Operation
EXPERIENCE 71-A 134
EXPERIENCE 71-B 420

72. Hunt Point Super-Market
EXPERIENCE 72-A 136
EXPERIENCE 72 B 423

73. Three Compressors Changed
EXPERIENCE 73-A 140
EXPERIENCE 73-B 424

74. Condensing Unit Location Change
EXPERIENCE 74-A 142
EXPERIENCE 74-B 427

75. Wrong Diagnostic-Water
EXPERIENCE 75-A 144
EXPERIENCE 75-B 429

76. Low-Pressure Controls Adjustment
EXPERIENCE 76-A 146
EXPERIENCE 76-B 430

77. Another Technician
EXPERIENCE 77-A 148
EXPERIENCE 77-B 431

78. Surplus
EXPERIENCE 78-A 150
EXPERIENCE 78-B 432

79. Cooling Tower Lines
EXPERIENCE 79-A 152
EXPERIENCE 79-B 433

80. Compressor's Start but Kicks Out
EXPERIENCE 80-A...154
EXPERIENCE 80-B...436

81. Low-Pressure Control Adjustment
EXPERIENCE 81-A 156
EXPERIENCE 81-B..438

82. Fish Tank Refrigeration System
EXPERIENCE 82-A 158
EXPERIENCE 82-B 439

83. Low-Pressure Control Installation
EXPERIENCE 83-A... 160
EXPERIENCE 83-B 440

84. Semi hermetic Unit Running with a Higher Voltage
EXPERIENCE 84-A...162
EXPERIENCE 84-B 441

85. Sudden Defrost—Back to Frost
EXPERIENCE 85-A... 164
EXPERIENCE 85-B 443

86. Chiller Coil
EXPERIENCE 86-A... 166
EXPERIENCE 86-B 445

87. Appropriate Use of Electrical Tools
EXPERIENCE 87-A 167
EXPERIENCE 87-B 447

88. Repair of a 15-Ton A/C System
EXPERIENCE 88-A...169
EXPERIENCE 88-B 450

89. Testing Commercial Equipment for Leaks
   EXPERIENCE 89-A..171
   EXPERIENCE 89-B..453

90. Butchery-Meat Cooler
   EXPERIENCE 90-A..173
   EXPERIENCE 90-B.. 454

91. Refrigerant Leak in a 100-Ton Semi hermetic Compressor
   EXPERIENCE 91-A 175
   EXPERIENCE 91-B.. 456

92. Changing an A/C Semi hermetic Compressor
   EXPERIENCE 92-A... 177
   EXPERIENCE 92-B 458

93. Water Bill Too High
   EXPERIENCE 93-A..179
   EXPERIENCE 93-B 460

94. A/C Equipment Does Not Work
   EXPERIENCE 94-A..181
   EXPERIENCE 94-B 462

95. Refrigeration System to Keep Corpses
   EXPERIENCE 95-A..183
   EXPERIENCE 95-B 463

96. A/C Installation Check (School)
   EXPERIENCE 96-A..185
   EXPERIENCE 96-B 466

97. Freezer's Unit Burned Out
   EXPERIENCE 97-A 188
   EXPERIENCE 97-B 468

98. Refrigerant Leaking
EXPERIENCE 98-A 190
EXPERIENCE 98-B 470

99. Solenoid Valve Coil
EXPERIENCE 99-A 192
EXPERIENCE 99-B 472

100. RSES Seminars
EXPERIENCE 100-A 194
EXPERIENCE 100-B 474

101. Medium Temperature Walking Box
EXPERIENCE 101-A 196
EXPERIENCE 101-B 475

102. Water vs. Glycol
EXPERIENCE 102-A 198
EXPERIENCE 102-B 477

103. A/C Operation
EXPERIENCE 103-A 200
EXPERIENCE 103-B 478

104. Medium Temperature Cold Cuts Spoiled
EXPERIENCE 104-A 203
EXPERIENCE 104-B 479

105. Poor Cooling Effect
EXPERIENCE 105-A 205
EXPERIENCE 105-B 480

106. Ice Maker—Harvesting Problems
EXPERIENCE 106-A 207
EXPERIENCE 106-B 482

107. Air Conditioning—Poor Cooling
   EXPERIENCE 107-A...210
   EXPERIENCE 107-B.. 483

108. Ice Maker—Operational Problems
   EXPERIENCE 108-A 212
   EXPERIENCE 108-B 484

109. Fan Motor Rotation
   EXPERIENCE 109-A 215
   EXPERIENCE 109-B 485

110. Installation Care
   EXPERIENCE 110-A 217
   EXPERIENCE 110-B.. 486

111. Instruction
   EXPERIENCE 111-A...219
   EXPERIENCE 111-B.. 487

112. Thought: To Want Is to Be Able
   EXPERIENCE 112-A 222
   EXPERIENCE 112-B.. 489

113. My Affiliation with RSES
   EXPERIENCE 113-A 225
   EXPERIENCE 113-B.. 490

114. Installation Error? (Connecticut)
   EXPERIENCE 114-A.. 226
   EXPERIENCE 114-B..491

115. Reunion with Former Student
   EXPERIENCE 115-A.. 228
   EXPERIENCE 115-B.. 492

116. Supermarket: Flooded Cellar
    EXPERIENCE 116-A 230
    EXPERIENCE 116-B.. 493

117. A/C Problem
    EXPERIENCE 117-A..231
    EXPERIENCE 117-B 494

118. Pollution Certificate
    EXPERIENCE 118-A.. 233
    EXPERIENCE 118-B 498

119. Heating Problem
    EXPERIENCE 119-A.. 235
    EXPERIENCE 119-B.. 499

120. Man from Vegas Course of A/C
    EXPERIENCE 120-A.. 237
    EXPERIENCE 120-B 500

121. Partner—Electric Diagram—System
    EXPERIENCE 121-A.. 238
    EXPERIENCE 121-B.. 501

122. Water Dripping From Condenser's Pan
    EXPERIENCE 122-A 240
    EXPERIENCE 122-B 503

123. Irish Coffee Pub
    EXPERIENCE 123-A 242
    EXPERIENCE 123-B 505

124. Excessive Cost of a Hammer
    EXPERIENCE 124-A 244
    EXPERIENCE 124-B 506

125. Electrical Problem with the Car
EXPERIENCE 125-A 247
EXPERIENCE 125-B 507

126. Leaking Refrigerant
EXPERIENCE 126-A 250
EXPERIENCE 126-B 509

127. Compressor's Parts Leaking
EXPERIENCE 127-A 252
EXPERIENCE 127-B..511

128. Medium Temperature Frozen Products
EXPERIENCE 128-A..253
EXPERIENCE 128-B 514

129. Ice Blocks Calculations
EXPERIENCE 129-A..255
EXPERIENCE 129-B 516

130. Faulty Compressor
EXPERIENCE 130-A 257
EXPERIENCE 130-B 518

My Railroad Work's Experience - 1 260
My Railroad Work's Experience - 2 265
My HVAC & R teaching beginnings 269
"More than just R and A/C" Experiences................................ 273
Other School Teaching Experiences.. 275

Formulae & Other Calculations..521

Introduction

HVAC & R Hands-On Troubleshooting

By the time I decided to start writing this book, I had worked in the heating, ventilation, air conditioning, and refrigeration (HVAC & R) industry for more than forty-five years. In this span of time, I had worked as a service technician in an NYC service company, and as a trade instructor in several schools.

I have written books and works, among them refrigeration, air-conditioning, and electric (RAC & E) test books, preparation for the Environmental Protection Agency (EPA) "Refrigerants usage certification," and the Refrigerating Machine Operator (RMO) License for the NYC Fire Department, which had been used in the schools in which I worked.

Regardless of the years that had passed, the refrigeration system used in air conditioning systems as well as in commercial refrigeration, domestic refrigeration, and in the equipment in general used today contains the same mechanical-electrical components as then.

For the above reasons, I hope my experiences here related could be used to troubleshoot, to diagnose, and to serve any one of these systems.

I know that *all* the methods I used in the past are without any doubt the same ones being used today. Of course, each method (procedure) could be varied (considering the changes that have appeared in the HVAC & R industries)—tools, components, oils, refrigerants, etc.—in different forms, as it was used then.

The problems found then (with some small variations) are the same ones that can be found today. Therefore, the same remedies, changes, and repairs used then can continue being used today.

I am aware that nobody wants to learn by the experience of others, but anyhow I think—hope—that by reading these experiences, a reader may find valuable knowledge that can be applied to make some advances in his or her trade.

Perhaps the information given here may not be important to everyone, and may create in some people controversy, but this is a recounting of my HVAC & R working-teaching experiences. So according to these, let us try something:

In every one of these short service histories and experiences, I will *take you* (the reader) with me so we can start applying our knowledge and together we will troubleshoot every system or complaint. See Experience 32.

The experiences written here are presented in two segments, "A" and "B":

"A"

    a)   *Complaint, according to the dispatcher's information.*
    b)   *Information provided by the customer or person in charge of the equipment.*
    c)   *Equipment's mechanical, electrical, or other characteristics.*
    d)   *Others.*

Some questions related to points a, b, c, and d above.

"B"

    a)   *Analysis of complaint.*
    b)   *Calculations, when necessary to make clear the problem and solution.*
    c)   *Diagnosis of the problem.*
    d)   *Procedures carried out to verify the problem.*
    e)   *Solution to the problem.*

f) *Recommendations, when necessary.*
g) *Others.*

<u>Thought:</u> I always knew, and I had no doubt, that we are thinking humans, *not* robots.

For this reason, whenever we were, or are ordered to perform some work-job, we may have the right to verify what we were told, in order to reach a perfect job's end!

This, in a short or long time, will give everyone a clearer idea of what we are doing or what we are trying to do.

In every one of the experiences related here (and many of them seem unbelievable), I only limited myself in applying my knowledge and experience as I learned, and through it found, the problem. In addition, in every one of the experiences, I express what I had observed and what the customer told me.

In any service call, there are three important steps that we must complete to overcome the problem or complaint.

<u>First step.</u> Compile from the customer or person in charge of the equipment as much information as possible.

He or she may give us the insides and history of the problem or complaint.

<u>Second step.</u> Familiarize yourself with the electrical, mechanical, and general characteristics of the equipment.

<u>Third step.</u> Be aware of what the system or components are supposed to do and what they are doing at the time of service.

Usually, I believe the system or component tries its best to inform us of the trouble. The problem is that sometimes we use the wrong "language" with the system/component and we do not understand it.

Then, by using the above information, our common sense, the appropriate tools, the procedures, the electrical meters, etc. we can appropriately diagnose and fix the problem/complaint.

Collected here, there are one hundred and thirty service experiences; they appear, in no chronological order. This work has being elaborated with the idea, of encouraging the use of appropriate procedures in the service of these technical branches.

With the purpose of obtaining, through the reading of this book, the greatest possible benefit, please read separately each experience. First part "A," then part "B" of that experience.

To facilitate the understanding of some A/C & R systems operations; beginning on page # 521 we have shown some Formulae and other calculations.

A Refrigeration Experience

In the beginning, it was more wondering than interest.

I believe it was in the middle of February. Once the ship's propeller repair was performed (a problem happened on the last trip), we were prepared to sail. Like it was customary in the morning of the day previous to the trip, the refrigeration system was started up in order to get it ready to receive the meat, vegetables, and other foods requiring refrigeration. The foods were expected to arrive at about 6:00 or 7:00 p.m. I was the electrician on duty. Among other functions, I had to take hourly data of the refrigeration system (operating pressures and temperatures). To accomplish it, I didn't have to know anything about refrigeration.

At this time the head of the electrical department (in charge of the equipment) was a master chief with the last name Moreno. This particular day, he went on liberty at 4:30 in the afternoon. At about 5:00 p.m., I went to the refrigeration machine room to fulfill my duty and I found that the system it was not operating and temperatures were rising. I didn't know what to do, so I went to the gangway to report the problem to the officer (a lieutenant) on duty.

In view of such emergency, the officer called a police patrol and gave them the order to go and look for chief Moreno. It seems that he had said where he was going to be, because the police on patrol went directly to a brothel at the tolerance district (Tesca); there they found him, in between two women, embracing them both.

I was pending—awaiting the chief's arrival. When that happened, he (with a mocking smile) asked me, "What happened, couldn't repair the refrigeration system?" I obviously answered no.

I followed him to the refrigeration machine room, eager to know what had caused the problem. When we arrived there, he asked me for a screwdriver; I left *flying* to bring it; when I returned to the machines room, the equipment was already working. I asked him what had been the problem, and (as always, with the mocking smile on his lips and a lot of sarcasm) he answered me, "Someday you will know what was the problem."

Then, I suspected—but some time later, I was sure—that the chief Moreno "prepared" the problem, so that everything happened as it did. Years later, I had him as a refrigeration instructor. Truthfully, he never convinced me about his knowledge in R and A/C.

How My Interest in Refrigeration Started

In this year (1958), in the middle of the month of June, we weighed anchors again, this time, in Panama's direction. We passed the channel and we went south, once again to Buenaventura, a Colombian port in the Pacific Ocean.

In this trip, it happened, or better, it began—perhaps the most important experience or stage of my life.

The ship *Admiral Padilla* (a frigate), where I was assigned, was supplied-prepared (as it was customary for a long trip). By that effect, were stored in the two refrigeration rooms.

In the low-temperature room (freezer), there were eight heads of cattle (approximately six thousand pounds of meat), and in the other room, at medium temperature, were the rest of foods requiring refrigeration.

We left port, and everything went well until the fifth day of the trip (according to my calculations, about 1,500 miles from Buenaventura) when the refrigeration system broke down. Nobody in the electrical department who was in charge of this equipment or in the entire ship knew how to fix it. Then, I did not know anything about refrigeration.

Then I asked myself how it was possible that in a ship with a crew of more than two hundred men, nobody knew anything about fixing this piece of equipment.

It gave me pain to see that thousands of pounds of meat were hurled by the hut because the meat already began to get spoiled and of course to smell.

The captain decided to interrupt the trip. We returned to Panama; there, the ship was taken to the shipyard to see if they could repair the refrigeration system. I guess that according to the received information, the commander preferred to return to Cartagena port and to repair it in the shipyards there.

Throughout the return trip, the first thing that I did was to go to the ship's library to look for and to begin to read all available books and information related to refrigeration, or anything that I could find in the ship's bookstore.

From then on, I spent whole hours, with a dictionary's help, translating from English to Spanish (*The U.S. Navy's Manual, Chapter 59)* and, with the obtained information in the refrigeration equipment room, watching and trying to figure out not only the system's operation but also each of its mechanical-electrical components.

I believe that learning the refrigeration system operation became an obsession to me; I put myself (literally) inside the system, within the pipes, components, etc. I tried to figure out what happened within to the refrigerant in each one of the system's mechanical-electrical components.

This was the way I started seriously getting interested in this trade.

And I had to do it all by myself, because I did not have anybody to ask about it. I put into practice all the mechanical knowledge that I acquired in St. Antony (my youth school) as well as in my five years' work in the railroads, in addition to my good common sense. It was as much my desire to know the operation of the system so that everybody in the ship realized it and that when they needed me, they knew where they could find me.

I came to understand something about the system's operation, but I knew that I needed much more. I believe that the information that I found on board was good but under no point of view sufficient.

===

EXPERIENCE 01-A

On the last days that I served in the Colombian Navy, I remember some incidents. The one related here was very important to me:

The U.S. Navy sent three Coast Guard vessels to serve in Hawaii. These ships sailed from Mobile, Alabama, and each ship was equipped with a twenty-five-ton air-conditioning system. Two of these systems did not work. The ships were taken to Coco Solo, Panama's shipyard, where they determined that the machines had refrigerant leaks, but these leaks could not be located. Because it was proximate, the vessel's captain communicated with the U.S. naval mission in Barranquilla, Colombia.

They could take the boats back to Mobile, Alabama, or they could take the boats to this city, which was much closer. In the naval mission, Lieutenant Johnson decided to call me to troubleshoot the problem. I remembered my old man always said, "A good soldier neither offers nor refuses." I decided to do my best and prepared some questions for him to ask the captains of both ships. From their response, I had an idea of what the problem was. I explained to him the procedure that could be carried out. Both the captains and Lieutenant Johnson agreed with my plan.

I had taken some training courses in the USA, and was thankful for my education; I saw this as my opportunity to give back something from that knowledge. They took me to the ships to carry out an inspection. As we were down to the lower decks of the ship, I remember bringing some leak-detection tools. I was introduced to both captains and both engine room chiefs, who showed me the two systems that were leaking.

Note. Since the two vessels were of the same type, and used the same R-12, the exact same leak testing procedure was to be performed in both of them.

Before starting any procedure, we also verified the "lead fingers" (four in total) located at the condenser's heads. These pieces of metal are used to minimize the corrosion effect caused on the condenser's metals by the galvanic currents induced by water friction when flowing through the condenser. They were found at about 85% of their efficiency.

Questions

1. If you were called for a problem similar to this, what would your idea be as to where the refrigerant's leak is?

2. What question should you ask the captain or the engine room chief?

3. What would be the first procedure you should perform?

4. What tools do you need to carry out the procedure?

5. What should the test pressures be?

6. What should be the operating pressures in the system?

===

EXPERIENCE 02-A

This experience was relevant in my life after leaving the Colombian Navy. In this experience, as in other experiences in the future, we should study the complaint or problem, and through the analysis and applying our knowledge, the system should be repaired.

Somebody in the naval base told me that in the cafeteria this particular beverage cooler (property of the Beverages Factory) used to promote its products and lend their customers these commercial refrigeration units, which weren't working well. In those days, the beverage company had a private refrigeration service company to take care of the maintenance and repair of those units.

According to the person in charge, for the last several days the device was not cooling properly. The mechanics they sent all had problems in repairing the unit. He enumerated to me what the mechanics did on the three visits:

1. The first mechanic's visit was due, of course, to a poor cooling of the device. He installed the gauges and "worked" in the equipment for a while. After about fifteen minutes he said the system was OK. And he left.

2. The equipment continued with the same problem. The following day he called the beverage company and another mechanic showed up. This time the mechanic worked a little bit longer and finally said the same as the other mechanic, then left.

3. On the third day, the equipment still failed to work properly. He called the beverage company again. And another mechanic was sent. This one connected the manifold gauges and said that he added some refrigerant and made some adjustments.

This time the cooler worked a little bit better, but it was not enough.

The system had the following characteristics:

a. It used a 1/3-hp air-cooled hermetically sealed condensing unit, equipped with a packing gland suction service valve attached to the compressor.
b. It used a 1/9-hp condenser fan motor.
c. Low-pressure control. Actual settings = Cut-in. 35 psig; Differential = 25 psig (*)
d. Both motors (working), compressor and condenser fan, used the same electrical characteristics: 115 volts, single phase, 60 Hz.
e. The system used refrigerant 12.
f. It used a capillary tube as a liquid control.
g. Ambient temperature at the equipment location: 80°F.
h. The FLA, discharge, and suction operating pressures, as well as the suction line and the liquid line's temperatures, were a bit too high.

The beverages required a temperature of 40°F.

Let's see what we can find.

Questions

1. What do you think the problem was?

2. What are supposed to be the operating pressures?

3. What should we do?

==

EXPERIENCE 03-A

This experience took place sometime later than Experience 02. The lieutenant aggregate of the naval mission was transferred back to the United States. Thus, he had to get rid of some things, among them an 18,000 Btu Carrier air conditioner. He had the bad luck of the equipment failing to work when he gave it to the new owner.

A/C's characteristics:

a. It was a 1 1/2-hp sealed rotary-type compressor.
b. It used a 1/2-hp, three-speed, double-shaft fan motor.
c. Selector switch. Thermostat. Thermal overload.
d. Both motors, compressor and condenser fan, were of the PSC type. Both had the same electrical characteristics: 230 volts, single phase, 60 Hz.
e. Used refrigerant 22.

The lieutenant got in touch with me and told me the problem; he asked me if I could check the device. Thus:

Questions

1. What do you think the problem was?

2. What test or tests should you carry out to verify this equipment?

3. What should you recommend?

===

EXPERIENCE 04-A

I continued working with the beverage factory (cited on Experience 02) without any problems. Being aware that I needed some spare replacement parts, I talked to the company's manager and suggested he buy several spare parts, which would be needed for stock instead of ordering one at each time. He did it, and stored them in the factory's warehouse.

On this occasion, I had one problem; when only two compressors were left in the warehouse, the person in charge suggested to me to remove them, and hold them in my shop. I accepted and took them with me. I used one of them immediately and later on (about two months), when I installed the other, and I went to start it, it did not work.

Motor's characteristics: 115 volts, single phase, 60 Hz.

Questions

1. What do you think the problem was?

2. What test or tests should you carry out to verify the problem?

3. What would you recommend?

===

EXPERIENCE 05-A

While working for a "Colombian Technical" Company ——, one of the various works that I remember was in Santa Marta City, on a 10-ton A/C system that had been operating for several years in the Clinic of Social Insurances. According to the person in charge, the equipment had lost little by little its cooling efficiency. For a while now, it's getting to the point where the evaporator, regardless of what they did (*), it was "frozen" all the time.

(*) According to the building's service engineer, in a span of about four years, these are some of the works that they and outside service companies performed on this A/C equipment:

1. Cleaned the evaporator with steam on a monthly basis.
2. Changed air filters on a weekly basis.
3. Changed the fan-blower to increase airflow through the evaporator.
4. Changed the evaporator fan motor (direct drive) with one of a higher speed.
5. Changed the evaporator fan motor type for an indirect drive (belts-pulley-flywheel).
6. Remove the refrigerant from the system and perform a deep vacuum to "clean" the system internally. This last procedure was performed several times.

The system had the following characteristics:

a. A self-contained system with a 10-ton semi hermetic unit, equipped with suction-discharge packing gland service valves type, attached to the compressor.

b.   Shell and tube water-cooled-type condenser.

Water from cooling tower, temperature 80°F.

c.   Used a 5-hp cooling tower fan motor.
d.   Controls: High and low pressure, water-regulating valve. Pump-down (thermostat—liquid line solenoid valve. Thermostat's sensing bulb located at the evaporator's return airflow).
e.   The evaporator's fan motor was of 5 hp. All the motors and the compressor, cooling tower, and evaporator fan had the same electrical characteristics 440 volts, three phases, 60 Hz.
f.   The system used refrigerant 22.
g.   The thermostatic expansion valve was of the external equalizer type.
h.   The sight glass in the liquid line was full.

The system's operation was observed and the following were all lower than normal:

1.   The suction and discharge pressures.
2.   The suction and liquid line temperatures.
3.   The running amperage.
4.   The condenser's water temperature difference.

After hearing the detailed explanation from the person in charge,

---

Questions

1. What do you think the problem may be? Keep in mind the system is old and money is not in abundance in this place.

2. What are supposed to be the operating pressures?

3. What are supposed to be the suction and liquid lines temperatures?

4. Why is the running amperage so low?

5. What should be our recommendations?

==

EXPERIENCE 06-A

At one time, I performed some refrigeration work for a man who was retired from the navy. He had a restaurant called "La Canada" (The Gorge). The work to be performed was in a freezer. This was a self-contained unit, located in the pantry with the refrigeration unit positioned at the end of the cabinet.

The equipment had these characteristics:

1. It was a 1/2-hp air-cooled sealed condensing unit, equipped with a packing gland service valve attached to the compressor's suction side.
2. It used a 1/8-hp shaded pole-type condenser fan motor.
3. Both motors, compressor and condenser fan, used the same electrical characteristics: 115 volts, single phase, 60 Hz.
4. The system used refrigerant 12.
5. The device had a capillary tube as a liquid control.
6. Temperature at the equipment's location was about 85°F.
7. The frozen products to be stored required a temperature of 10°F.

The freezer's compressor, although working, was not cooling, and if touched it "passed current." The equipment was de-energized, the entire electrical installation was reviewed and although the compressor worked, it was "grounded."

After installing the low side gauge, I found that the system did not have any refrigerant.

Questions

1. Up to this point, what do you think the problem was?

2. What tests should we recommend?

3. What do you think can be done?

4. What are supposed to be the operating pressures in this device?

==

EXPERIENCE 07-A

At this time, I had a call from a Panamanian ship's captain who said that he had heard of me and wanted me to inspect the refrigeration equipment on board his ship. He confided to me that he had gone to Panama's shipyards and the only thing that they had recommended was the purchase of a new compressor. That was a greater problem.

According to the captain, the compressor was of European origin. It was of the open "W" type, with 6 pistons and with a capacity of about 20 tons. Due to the compressor's age and origin, it was not easy to obtain.

Questions

1. What component parts should we recommend to be checked in this device?

2. What do you think can be done?

3. Any other suggestions to the captain?

4. Can we intent the compressor's repair?

===

EXPERIENCE 08-A

One day a retired sergeant major of the Colombian marines whose last name was Arias it came to my shop. He told me that he got some R and A/C works and asked me for help (in some way) to carry them out. He didn't have any mechanical experience.

The first work that he got was in the City of Cartagena, to repair a commercial refrigeration (walk-in box) used to store and freeze seafood products. The owner of this equipment bought the product from the fishermen. He stored it, and later he supplied it to the restaurants throughout the city.

We went to the place and inspected the equipment:

The repair was similar to one that I did on the boat (Experience 07); the only difference was that this equipment was on land.

The characteristics and requirements of the equipment or walk-in storage room were:

1. The compressor was an open "in-line" alternative type with 6 pistons.
2. It had a capacity of 25 tons.
3. The electric motor working at 440 volts, three phases, and 60 Hz. Approximately 550 rpm.
4. The system was working with R-12.
5. Thermostatic (external equalizer) expansion valve.
6. Required storage temperature was -10°F (-23°C)
7. High and low, separated pressure controls.
8. Condenser water-cooled "shell-tube" type (seawater). Water temperature was regularly approximately 80°F.

While inspecting the equipment, we found out (through the owner) that a former student of mine had been there, trying to make the repair. The peculiar thing was, according to the client, he tried to make the repair with the book in hand. In addition, he did not show that much experience, and it made the owner apprehensive as to the type of repair to be done.

After the inspection of the entire system and of all the spare parts, we reached at the conclusion that the work can be performed.

Questions

1. What should be the operating pressures for this system?

2. What tests should we recommend to make sure about the equipment's condition?

3. Should we use some of the information on Experience 07?

===

EXPERIENCE 09-A

At this time, I met with my brother-in law (he was working for an A/C company), and I asked him what he was doing. He said that he was working with a mechanic in a bank and there, they had some problems. The compressor was repaired. Later, it began to present several compressor faults (rupture of connecting rods, internal valves, and valves' plates, etc.).

Since we were very near the bank, I decided to go there and look at the problem. When we arrived I met with the mechanic. After he answered my greeting, not happily (I didn't have a good relationship with him), I asked him about the problem and he told me that the same problem persisted, the refrigerant returning to the compressor in a liquid state.

He was frank and told me that he didn't understand why.

Most important equipment information:

1. Alternative open-type machine of 150 tons.
2. With a liquid control of the pilot expansion valve type.
3. This system used R-22.

In the operation of this system, a small thermostatic expansion valve is used, hence the name. When the superheated vapor refrigerant at the evaporator outlet sends the signal to the valve to open, the liquid refrigerant passing through the pilot valve activates (overcomes) a spring force placed on the piston to keep it closed. The mechanism attached to the piston (of about 3.5" or 4" diameter) moves a seat of the main valve, which allows the liquid refrigerant to flow into the evaporator.

Questions

According to the operating method of this device,

1. Why is the liquid control allowing the refrigerant liquid to return in that state to the compressor?

2. How do you think the closing movement of the main valve is accomplished?

3. What should we think the problem might be?

4. What should we inspect to be able to solve this problem?

==

EXPERIENCE 10-A

This service call was for a Domestic Refrigeration application, a household "Refrigerator" with one door. Particularly, the client's complaint was "The refrigerator, works well, then suddenly defrosts, remains in this condition until all the ice on the evaporator surface disappears; then it returns to freeze again." The condition repeats itself the same way, over and over again.

This system had the following characteristics:

a. It used a 1/8 hp sealed unit.
b. It used a static-type condenser on the back of the box.
c. The compressor's motor (working) had the following electrical features: 115 volts, single phase, 60 Hz, 1.8 amps.
d. The system used refrigerant 12 as stated on the ID plate on the back of the box.
e. Ambient temperature at the equipment location: 80°F.

Notes: Non-automatic defrosts. No fan.

As usual, the lower temperature requirement in this type of equipment was 32°F (ice making).

Moreover, according to the client, this refrigerator was a gift from her son about five years ago. Until one week ago, the device never presented any problem.

Questions

1. What do you think the problem was?

2. What should you do to verify this problem?

3. What explanation should you give to the customer?

4. In normal conditions, in this type of equipment, besides the compressor's noise; what other noises should we expect to hear from a normal operating system?

===

EXPERIENCE 11-A

We were overseas, and as always, there was the opportunity to learn something. We were in Stockholm, Sweden. The ship's commander gave permission to the married guys to buy some appliances for their home's kitchen. I went with a crew member and a friend of mine who wanted to buy a refrigerator. At the supply house, we met very amiable people with the spirit to help. The language was completely different, Swedish, but many people spoke English and more or less we could communicate to each other.

The employee asked us where we were going to take the refrigerator. We answered to Colombia, South America. He asked specifically where in the country, and we answered Barranquilla (average temperature approximately 85°F, or 29°C.). He said to us that in that case, the refrigerator had to be "tropicalized"; that is, that if we brought it as it was, it was not going to work normally. He said to us that they had to prepare it for the new "condensing" temperature where, it was going to work. He said to us that completing that procedure required a couple of days.

He gave us their telephone number, and the way in which we could communicate with him when we returned, if the ship used the same route. Then we would let him know to do whatever was required.

Note: The term "tropicalization" then, as it implies now, has to do with the condenser size. It means to prepare the equipment to work properly in the tropic or at a different ambient temperature. It had to be resized for the new temperature. It is the same thing as if a refrigerator working in New York City (kitchen's average temperature +/-70°F [21°C]) were relocated to another temperature, or at a higher or lower ambient temperature such as Miami or Boston. It would never work right in another location.

The condenser's size could be either too small (Miami) or too big (Boston).

Questions

1. Do you think that is important, the size of the condenser for the system's normal operation?

2. What is your opinion?

3. What would you do in case the condenser was too small?

4. Do you have any other ideas on how to solve this problem?

5. What should be the size of the condenser if the refrigerator had a capacity of 1/3 hp?

===

EXPERIENCE 12-A

This experience happened overseas, but its learning effect can still be used in these days, and it may be applied anywhere because its characteristics can be found at any time and at any place. This company had a service refrigeration mechanic, and by from his information I found out that they had several "abandoned" household refrigerators in the company's backyard. He told me that several months back, they had imported from the United States twenty-five of those refrigerators. They were made by ——.

The company sold them to the public, but none of them worked at all, and the company left them as "lost." The mechanic suggested to me that I offer some money to buy them and perhaps I could repair them, and get some benefit from that transaction.

Before making any offer, I went to the company's backyard to see the appliances. I checked them. Inside, they were brand-new (they even had the ice cube moldings, in the original packaging) but externally they were in very bad shape. They were corroded, and the sheet metal needed to be repaired and painted. Anyway, I chose six of them (the ones in better shape) that I believed I could do something with. I do not remember how much I paid for those refrigerators.

Here, I would like to emphasize once more that it's very important in any case to get familiar (with *all* the electrical as well as the mechanical characteristics) of the device or equipment every time you are called to work or service them.

The characteristics of these machines were as follows:

1. Their capacities ranging between 1/8 and 1/3 hp using relays of the current type.

2. Most of them (1/3) had two side-by-side doors; the 1/8 only one door.
3. All of them had the same electrical characteristics: 115 volts, single phase, 60 Hz.
4. Condenser devices were located on the back of the refrigerator.

Anyway, I took, the refrigerators to my shop and by using the appropriate electrical instruments I verified them. All the electrical resistance readings (ohms and ground) were OK.

The ambient temperature at my shop location was at this time about 80°F.

Therefore, I energized one of them. I checked the starting, then the running amperage. Electrically it ran well, but it did not show a cooling effect at all.

Questions

1. What may we think the problem was?

2. How should we verify the refrigeration system operation in equipment like these?

3. Other than the compressor running, when a refrigeration system like this starts to work, what noise should we expect to hear?

4. What temperatures should we expect to feel on the system connecting lines?

5. What investigation or procedure should we conduct?

==

EXPERIENCE 13-A

This time, I was consulted for a problem in an industrial refrigerator (walk-in box) located at the municipal slaughterhouse in Barranquilla, Colombia. The problem was that some of the stored meat got rotten.

To provide the cooling effect, the refrigeration system used ammonia to chill brine (sodium chloride or NaCl), the secondary refrigerant, which by means of pumps was delivered to the top of a fountain-like structure, positioned systematically in a huge refrigerated space. Brine descended on eight wooden slats staggered horizontally, approximately six feet long by six inches wide, and was collected in tanks at the bottom of the slats, then back again to the cooling tanks.

The problem was not with the cooling system, but because of the size and importance of the system, I decide to take advantage of the opportunity to ask some questions and look at the entire equipment.

System Characteristics

1. Alternative open-type compressor of several hundred tons.
2. The electric motor used to move the compressor, worked with 440 volts, three phases, and 60 Hz.

    Note: The entire electrical power supply for all the motors had the same characteristics.

3. A huge cooling tower.
4. Two sets of pumps:

    (1) One pump was used for the condenser's water circulation (temperature about 85°F.).

(2) The other pump was for the brine circulation (supply and return lines insulated).

Evaporators immersed in a brine-tank about 60' x 60' x 20' (about 72,000 cu ft).

Temperatures of equipment operation were thus:

Boiling point ammonia (R-717) = -15°F. (-26°C.)
Brine temperature = 0°F.
Brine density = 1.164 (g/cm3)
Space temperature = 15°F. (meat)

In this place, the newly slaughtered cattle meat was stored on "hanging" hooks, which in turn ran on rails fixed to the ceiling. When the meat was about to be placed on the delivery trucks, it was easily moved to the storage doors.

Questions

1. What is the problem?

2. What should we do to find the object of this complaint?

3. What should we think were the system operating pressures?

4. Should we have any other questions to be asked?

==

EXPERIENCE 14-A

As mentioned before and more than ever, when working with any refrigeration or air-conditioning system, we should be very careful not to mix refrigerants. Not only that, its compatibility with any other chemical substances being used within the system (oil, tracers, etc.) should also be checked. This story happened outside the United States, but in today's world, and as quoted in Experience 15, this problem can occur anywhere.

Then, there were some refrigerants of unknown source. A lot of refrigeration and air-conditioning service men were using them, not that they were better, but they were less expensive than others known (up to 50% cheaper). Because we were facing a lot of problems with, I requested seller's literature on those chemicals. When reading it, I saw something that I didn't like, and that gave me some distrust. According to the manufacturer's once oils were exposed to them; shouldn't be bare handled because they could cause fungus on the fingers. According to my knowledge and experience, the refrigerants (used then) were so safe that food could be exposed to them and did not cause physical harm of any kind.

At that time, it presented a burning "epidemic" of sealed or semi sealed units (motor-compressor). As I performed it on many occasions, I opened several units to carry out the rewinding of the electric motors in them. The contamination was so intense (the oil appeared as "mud"), I had to intensify efforts in cleaning and decontaminating the system.

Someone who had experience with the same problem suggested to me to "wash" with water the whole motor-compressor unit. Therefore, I did it, and the water was able to remove all the pollution, mud, etc., quickly and effectively.

This action was not possible to be performed with the rest of the system (tubing components).

Questions

1. What should we do to assure the compatibility of the substances we use?

2. After a unit burnout, what are your suggestions for decontamination of a system?

   a. Can you describe a decontamination process?
   b. What determines point "a"?

3. What test should be performed to verify how serious the contamination in a system is?

===

EXPERIENCE 15-A

This service call was to an ice blocks factory. Once the primary refrigerant (ammonia) cools the brine (calcium chloride), it is sent to the water-containers tank.

According to the customer, the system was not cooling well. Up to this day, he said, two other companies have reviewed the system, and found no problem.

The system had the following features:

References

1. It used a huge, open-type compressor. It worked with refrigerant ammonia (NH3) (R-717)
2. Condensation by water (cooling tower) at 80°F. water temperature. (Δt - Dt = 10 to 12°F.)

Requirements

1. Ice blocks temperature   = 18°F (-7.7°C)
2. Refrigerant boiling point = -12°F (-24°C)

Normal operation pressures
High = 215/245 psig (105/115°F)
Low = 7.8 psig (BP -12°F)

Note: This equipment had the operating pressure gauges installed on a board at the wall.

According to calculations noticed:

(1) Both operating pressures (discharge 260; suction 12 psig) *too high.*
(2) Water temperature difference between the supply and return (25°F.) *too high.*
(3) Water returning to the cooling tower was *scarce and too hot.*
(4) Water pressure (approximately 13 psig) *too low.*

Questions

1. What do you think the problem may be?

2. What kind of test (or tests) should we perform to find out the problem?

3. What should be the amount of water gallons per minute (gpm) through the condenser for this system?

===

EXPERIENCE 16-A

I came to New York City with a working contract, hired by this RAC service company. On the first day, after being introduced and filling and signing papers and other requirements, I was ready to start to work. Since this day, every morning I waited by the dispatcher's window until he informed me with whom I was going to work every day.

Beginning on this day, in every job I was involved, I tried to apply and show my knowledge and experience. They started noticing that I knew what I was doing.

I always had a great illusion to work in the USA, and to learn the techniques I believed they had in this country. However, from the very beginning I experienced problems, one of them being my way of working. I had to learn not too well that here (many times) some R & A/C mechanics do not make repairs instead they replace (many times unnecessarily) equipment and parts.

Another problem I had was that I never carried a toolbox with me; I only wore what I believed was necessary (today, I think the same way): my head (common sense), a screwdriver, a lantern, a pair of pliers (in my shirt's pocket), a PT chart, a thermometer, and, as I believe that at least 70% of service calls are related to electricity, a multi-meter.

While I was working in this service company (approximately five years), many events happened in my work. Some of those events were very interesting, to me anyway. Here, I shall try to enumerate them. Not chronological order. No names are given.

After all, more than forty years had passed.

It was summertime and approximately eleven o'clock in the morning. I do not remember very well, but I think this was one of the first "important" jobs that I attended as a helper in this company.

It was to a restaurant located at Lower Manhattan.

The system's evaporator was located at the back (dining room) of the first floor.

The problem was found almost immediately:

A lack of refrigerant, due to a refrigerant's leak at the flare nut on the TXV's inlet.

The mechanic tried to retighten the flare nut, but the leak continued.

The flaring on the tubing had to be remade.

I believe this mechanic already had been here before, because he showed a lot of familiarity with the system and the place. We went immediately to the roof. This building had six floors, so, we took the elevator till the sixth floor, and then we ascended the stairs to the roof.

In the middle of the place, there was a 5-ton (*) air-cooled condensing unit.

Then, following the mechanic's indications, I removed the covers to get access to the unit. To pumped down the system, and gather the refrigerant, he proceeded to close (front-seated) the service valve at the receiver's outlet. Then he said, "Let us go!"

Questions

1. What should be the first question to ask?

2. Should we make any suggestions to the mechanic?

3. What should be the first procedure we may perform here?

4. According to this system's capacity, how many pounds of refrigerant should we charge to it?

==

EXPERIENCE 17-A

This experience took place outside the USA, at a fairly large farm. The owner of the farm got a hold of me to ask if I could do something by setting up the equipment that would maintain the oranges fresh. He harvested, among other foods, about 15,000 oranges (approximately 10,500 pounds). By the time harvesting began and the delivery of the fruit to the warehouse started, a refrigeration system was needed to protect the fruit from spoilage. The distance to the warehouse from the orchards (about 10 kilometers or approximately 6 miles away) surely spoiled the fruit.

In addition to this problem, the owner of the farm was also contemplating exporting the fruit. So he asked me if I would be able to calculate and build a refrigerated box that would store the oranges. The fruits would be stored for approximately three weeks from harvest time, then sent to the warehouse.

The owner contracted me to do the job.

In order to do so, I thought, it was necessary to know the size of the space, the construction materials, the number of crates needed to store the product, and the size of each one of them.

Questions

1. What other items should we list to fulfill the customer's wishes?

2. What calculations could we perform to know the system's capacity?

3. Besides the product heat load, what other heat loads should I consider to accomplish the total heat load?

4. Do we have any other questions to the customer?

===

EXPERIENCE 18-A

Where this experience happened doesn't matter; it could happen anywhere, not necessarily using the same materials. In this case, it was a lacquer paint and rubber plastic (door gasket). Upon the customer's request, a painter in my company was sent to a residence to paint a domestic refrigerator.

He used white lacquer paint (from a local source). According to the client he had seen it in a warehouse of refrigeration parts, a light-blue-colored door gasket; he thought that the refrigerator would look very nice with such a combination of colors. To please the customer, I went to this warehouse and bought the gasket (foreign made).

We allowed the paint to set for four days. Past that time, we installed the gasket.

The next day the customer called, complaining the refrigerator's door couldn't open.

There was a chemical reaction between the paint and the rubber; they were melted—joined together.

We had to cut the gasket to open the refrigerator.

It isn't beside the point to say that we had to use the regular gasket type.

Questions

1. What should we do to prevent a problem like this?

2. Do you remember or had an experience of chemical reaction?

3. What did you do?

==

EXPERIENCE 19-A

This time I was sent to work with a mechanic to a bank on Columbus Avenue and 68th Street, it was about 9:00 a.m. The A/C equipment had some problems; the contactor for the last couple of days wasn't working well—it sparked when connected and disconnected. The mechanic left at about 9:30 a.m. saying that he was going to get a new contactor.

While he was absent, having nothing else to do, I disconnected the contactor and proceed to disassemble it.

The device's sets of contacts were cleaned with fine sandpaper, and I rearmed it and connected back to its working location. The A/C system was started. At the time the mechanic returned (approximately 11:30 a.m.), the equipment was working and the temperature of the place was OK.

Nevertheless, he said to me, that what I did was not the right way of fixing the problem.

Anyway, the contact was changed.

Questions

1. While the mechanic was absent, what else should you do?

2. Do you see anything wrong with this procedure?

===

EXPERIENCE 20-A

This day, I was assigned to work with another mechanic; he was a man of few words (at least with me) I thought it was a bitter man. I learned later, that he was getting divorced; and that he was running for supervisor. At the time, he was the shop steward. Somehow, this man looks like and reminds me of Hutch of the TV series *Starsky and Hutch* popular on those days.

In the company's office, he ordered me to go to the store and get five (R-22, 30 pounds each) refrigerant cylinders and put them in the truck (I thought, "This must be a big job"), and then he drove to a bank located at Broadway Avenue. I understood that he and another mechanic or supervisor had been working there for the last past two days.

I did not know, neither asked what work they had performed there. When we arrived at the bank, was about 8:30 a.m. and the Bank was still closed.

When the bank's guard opened the door and we entered, I saw in the center of the floor and against one of the columns an air-cooled A/C system. When the lower cover of the equipment was removed, I was able to see the equipment's characteristics:

1. A semi hermetic unit with a capacity of 5 tons.
2. The condenser, I assumed, was in some remote location.
3. On the left side of the equipment was a power supply box with three 30-ampere cartridge-type fuses.
4. Power supply was 230 volts, 3 phases, and 60 Hz.

Next, he asked me to bring all the five R-22 refrigerant cylinders, and the tools box along with the pressure gauges. *Without a vacuum or any other preparation to the system*, he just connected the manifold's suction hose to the suction's service valve, and began *adding the refrigerant, through the compressor's suction side, in a liquid state.* The compressor, responding to the abuse, started making all sorts of noises and within few minutes got very hot.

Questions

1. Do we see something wrong within this job?

2. So far, how many wrong things do you notice here?

3. Could you say something to the mechanic?

4. How many pounds of refrigerant should a system like this hold?

5. What should be the operating pressures in this system?

6. By working on this equipment for two days, don't you think we must know almost everything about this equipment?

After a while, seeing the system fail to operate properly, using all kinds of improper language, and kicking the compressor, he continued charging the refrigerant. And after adding almost three R-22 refrigerant tanks content (about 90 lb), he decided to leave.

==

EXPERIENCE 21-A

For a couple of days, I was "helping" the same mechanic as in Experience 20. This time we went to another place where two mechanics were working, these mechanics were in this location for several days, working in a 150-ton A/C system. While my boss was talking to the mechanic in charge, I talked to the other. I asked him if I could know what the problem was.

He told me that *regardless of what they did, the A/C's evaporator got frozen.* According to him, they had performed, for several times, everything they thought could produce such a problem. After cleaning repeatedly the TXV, they were cleaning the valve again. (The valve wasn't changed because it was too old and the customer expressed his intention at any time in the future to change the entire A/C system.)

Questions

1. Could you make a list of the reasons that A/C applications may induce a "frozen evaporator" condition?

2. From your list, do you have any suspicion or idea of what is making the evaporator in this system freeze?

3. What test, if any, could you carry out to repair such condition?

4. Do you have any recommendation to be applied here?

Let's remember that this problem is happening in an old A/C system and because we cannot assume the regular and usual "lack of refrigerant" or "dirty evaporator" or "dirty filters" or "something related with air flow."

We have to concentrate our attention on something else.

5.   According to the sentence above, do you have any other idea?

===

EXPERIENCE 22-A

On this day, I learned that some of the company's mechanics were working as instructors in trade schools such as Apex and TCI. They earned, I was told, about $5.00 per hour.

This day, I was assigned to work with one of those mechanics. We went to work on a rooftop. In past days, a mechanic had changed the cooling tower (5-hp fan motor). The motor was mounted with its base attached to the side of the cooling tower's structure.

Because of the way it was mounted, they realized that rain could fall into the motor and create a short circuit.

According to the mechanic's information, our work consisted in *lowering the motor and taking it back to the workshop to turn the covers so that water wouldn't penetrate.*

Questions

1. Do we have any suggestions about this job?

2. What should we think can be done?

3. Why wasn't the motor properly assembled at the shop?

EXPERIENCE 23-A

This was the same job and time as in Experience 22. I noticed that the surroundings of the cooling tower had an unusual amount of water. I asked the mechanic, and he told me moreover that this cooling tower, for a long time, was experiencing problems.

When the tower's *fan motor ran, it caused the spilling of large amounts of water.* That was the reason for the water on the roof; in addition, a (chemical) treatment for the water was spread, and as a result, a large number of dead pigeons appeared on the roof as well.

Questions

1. Do we have any explanation/suspicion of what is causing this spilling of water and the chemical?

2. Do we have any suggestions about something that we should do to end this problem?

3. Can we conduct some type of test or procedures that would enable us to identify the problem?

EXPERIENCE 24-A

For a while, I was helping a mechanic. He was a very good guy; he liked to drink beer, and in the mornings, almost always, he had a hangover. I liked to work with him because he always commented with me what we were going to do and asked me how I believed was my way of performing the job. Not only that, he always took into account my ideas.

In one of those works, I think it was the month of May, the dispatcher sent us to an immense computer room located on 45th Street between Madison and 6th Avenues to change a compressor.

We arrived at the place at about 9:00 a.m.

The compressor had the following characteristics:

1. A 75-hp semi hermetic unit.
2. The main power source from a three-phase circuit breaker, 440 volts, 60 Hz, and 3 phases.
3. This system used R-22.

Note: On the right side of this compressor, there was another one with the same characteristics. But I understood it was in a different electric circuit.

According to the person in charge, this compressor had already been changed twice at the end of the previous summer (because of a burnout).

Questions (besides the information given by the person in charge)

1. Do we have any other questions?

2. Do you think that we must investigate why the compressor had burned out so often before we change it again?

3. What do you think should happen if the compressor gets burned out again?

4. Should we have to wait until that happens again?

5. What procedures should be performed before we proceed to change the compressor?

I ask these questions because if this situation were repeated I shall feel responsible.

==

EXPERIENCE 25-A

This day, month of July of 1968, I was sent to help a mechanic whom was working at a restaurant. The dispatcher told me the mechanic was there for several days. When I arrived at the place, it was about 11:30 in the morning. I got into the restaurant (the place was very hot). I asked for the mechanic; they told me that I would find him around the corner, in the cellar.

I went to where they had indicated to me to enter the place. I lowered three steps and found myself in a space with a very low flat ceiling; so low (that it was necessary to walk bent). On my way in, I saw at least ten empty R-12 (30-pound) refrigerant cylinders.

At that moment, the mechanic was coming out (he was dirty and sweaty). Then he asked me, "What are you doing here"?

Then I was brand-new here. But after many years, still today, I have problems with that question (I can't decipher it). They asked the same question to me (many, many times), when I was sent to help somebody. I do not know if I were or had been wrong (I never want to think that it was something about discrimination). At the time, I interpreted it like they were saying I had nothing to do there. Besides, I am not going to deny it, but it bothered me since I was there to remove for them "the chestnuts out of the fire."

Anyway, after the "introduction" I asked the mechanic what seemed to be the problem.?

He told me that the equipment had a refrigerant leak. He handed me a halide torch (tool used to locate halogenated refrigerant leaks), and he said that he would come back. He went away to call the office.

I sat at one side of the equipment, and I began in getting familiar with it. This, it was a system with 50 tons capacity (I thought, no less than 200 pounds of refrigerant charge). The compressor was located on top of the water-cooled condenser-receiver tank (8' long and 3' in diameter).

I observed that in the discharge line of the compressor, they had recently made a strange "Y" connection: One of its legs came from the compressor, the other two went through a packing-gland service valve to the old condenser (below), and the third leg went to the new condenser. This was of the evaporative type; it was installed in another place of the cellar, at about a 35 feet distance from the compressor.

The compressor was stopped by the LP control; the pressure gauge on the wall showed a low-pressure reading (+/-20 pounds). The discharge pressure gauge reading was near the actual place temperature. Both valves, at the condenser inlet and outlet, were closed to the front-seat position. That meant, closed to the system.

Questions

1. With the collected information, what may we think?

2. In a space so close, if we were to use a refrigerant leak detector of any type, what do you think should be the leak detector tool's reaction?

3. Where do you think the refrigerant leak should be?

4. What tests may we conduct to find the answers to these questions?

5. Why were there so many empty R-12 refrigerant cylinders?

==

EXPERIENCE 26-A

I was sent to help a mechanic. As the dispatcher said, he was working for the last three days, in an A/C's cooling tower located in a fabrics factory, on the second floor of a building at 7th Avenue and 35th Street (the city's fashion district).

According to what I was told, the problem was that when the cooling tower's water pump stopped, the tower overflowed, flooding the place. To prevent this from happening, they must keep the cooling tower working at all times. This has been happening for the last week.

Once in the place, I started getting familiar with the system. The water piping from the cooling tower was connected to the water pump. From the water pump to the condensers, the supply line went vertically, then ran through the ceiling. The water-returning piping used the same running method and location.

Shut-off valves were installed at the water pump inlet and outlet.

Questions

1. What should be the first thing we do?

2. Do you have the idea why this flooding is happening?

3. Other than keeping the cooling tower working at all times, what else could we do to prevent this problem?

4. Why does it take so long to find and solve a problem like this?

==

EXPERIENCE 27-A

When finishing the previous service call (Experience 26), the mechanic called the office and it was told that along with me, went to a computers room located in 60th Street between Madison and 5th Avenues.

According to the computer room person in charge, several mechanics had been there, but they couldn't find the problem. This person told us as well that the building provided air-conditioning service from 7:00 a.m. until 6:00 p.m. After those hours, the place (computer room) had its own A/C equipment, the one that for the last several days did not work.

I suggested speaking with the building's engineer and trying to find something that may orient us (like a printing). The engineer, who has worked there for several years, said that in the building's printings, there was no existing information of such air-conditioning system arrangement.

We could not get any information.

Questions

1. With the above information in mind, what should we think the problem was?

2. Do we have any other questions to ask to the person in charge?

3. What can we do to put the system back in that automatic operation?

4. What or where do you think we should look for information?

5. What kind of device can this system use to accomplish that job?

6. Where can the control device be located?

===

EXPERIENCE 28-A

In order to prepare the air-conditioning equipment for the summer season, I was sent to this six-floor building equipped with A/C equipment of about 100 tons per floor.

Electrical Characteristics of the Entire Group of Systems

1. All the equipment worked with 230 volts, single/three phase and 60 Hertz.
2. A cooling tower (on the roof) provided the water to remove the heat from the condensers.
3. Most of the systems used R-22 and few of them used R-12.

The work to be performed included:

1. Changing all the A/C evaporator's air filters (approximately fifteen systems per floor, ranging from 3 to 7.5 tons).
2. Cleaning all evaporators.
3. Cleaning the cooling tower-condenser's water systems.

   Note: The water system (cooling tower, piping, and condensers) may be cleaned by using a chemical substance added to the cooling tower and circulated throughout the entire system.

4. Lubricating all the electric motor's ball bearings (cooling tower's water pump and fan, sump pumps, and evaporator's fan motors).
5. Verifying in all A/C systems the adjustment and operation of all temperature and pressure controls.
6. Verifying, changing, or adjusting fan belts.
7. Verifying the operating pressures, refrigerant charging, etc.

Do not rush, but do the job the best way possible.

Questions

1. How much time do you think this work would take?

2. What other tasks do you perform in a job like these?

3. Should we use a different method of cleaning the condenser water system?

===

EXPERIENCE 29-A

On another occasion, the same mechanic as in Experience 28 asked the dispatcher where I was working. He came to the place where I was and asked me to help him in a problem that he had with a 5-ton sealed A/C unit, which he was installing.

This A/C system had the following characteristics:

1. A water-cooled unit working with 230 volts, 3 phases, and 60 hertz.
2. A control circuit of 24 volts.
3. A sealed unit type with three-phase motor, windings protected with thermal overloads in each line.

I went with him to the place and checked the electrical installation. We found damage (burnout) in the three thermal overloads. The entire electrical installation was verified; nothing else was wrong. I suggested to him that the thermal overloads had come defective and to order new ones. And when the new thermal overloads arrived, I would help him to install them.

Two weeks later (when the parts came), I went with him and we made the change and finished the installation.

Questions

In an electrical circuit like this,

1. How many connections are there for thermal overloads?

2. Why were these thermal overloads damaged?

3. To prevent this from happening, what should we do?

4. On thermal overloads, what test should we perform to verify their condition?

5. From the test at 4, what information should we be looking for to decide their condition?

===

EXPERIENCE 30-A

It was about 1:00 p.m. This time, I was sent to help a mechanic who was working at the Carnegie delicatessen, located in NYC near the recital theater with that name.

The problem was that one of the food cooler machines on the store's main floor wasn't working properly and the mechanic could not find the matching condensing unit located at the basement, along with the other machines, about ten of them.

For some reason unknown to both of us the service company "forgot" to mark these systems appropriately. In addition, the mechanic does not have the slightest idea which system, out of the ten, to install the service gauges to service the specific faulty one.

Questions

1. What should we do to identify each of these machines?

2. From where can we start such endeavor?

3. What should be the logical way used to prevent the same problem in the future?

===

EXPERIENCE 31-A

The dispatcher sent me to help a mechanic whose last name was Rodriguez (in spite of his Spanish last name, he did not speak, or did not want to speak, even one Spanish word). I arrived at the place, a cellar, descended two steps, and I met with him.

He asked me the already usual question: "What are you doing here?" I gave him the usual answer: "I don't know; Larry just sent me over here to help you."

Anyhow, I asked him what the problem was. He a little reluctantly told me that there was a problem with the water pump's impeller. He said, "It had to be changed, but it was very difficult to remove it." Then he went away; I thought to bring a tool.

The impeller was almost completely destroyed, due to the effort made to remove it.

With the information above,

Questions

1. What should we do or suggest doing to remove this water pump's impeller?

2. What tool may we use to perform this job?

3. In the absence of the recommended tool, what other procedures should be used?

==

EXPERIENCE 32-A

In those days, they assigned to work with me an ex-accountant, who had withdrawn from Apex School. His name was Joe; I believe that he was in his early forties. According to him, he left the accounting business because of fatigue, and he was looking for a future in which he could work with his children (who were at that time still very young).

This man was very eager to learn, and asked an infinity of questions. He said to me that he had heard the other people in the company talk about me and that he was eager to work with me.

I thought, at least now I had someone with whom to express my ideas aloud. From then on, when I received a call from the dispatcher, I would communicate to Joe the information so he could analyze it, and he would tell me what he believed was the problem.

While we traveled from one place to another, he thought about the info that I had given him, and after evaluating it, he expressed his opinion to me. Due to his eagerness and hurry to gain experience, we did up to fifteen calls per day. He was very happy; I knew that he had asked Larry (the dispatcher) to please allow him to continue working with me.

Questions

1. What do you think of this working method?

2. Do you think this way of working is beneficial to someone?

3. Did you ever have any experience of this type?

Comment

As mentioned in the introduction, that's exactly what I'm trying to do now through these experiences.

Sharing information makes us able to widen our imagination and have a better idea of what to say and think.

===

EXPERIENCE 33-A

One morning (approximately 7:30), on a bright, sunny summer day, I called the dispatcher and he put me on the phone with a supervisor. He informed me that this time the problem was that for several years the evaporator of an air-conditioning unit located in a computer room always froze.

He enumerated to me what the company's mechanics had done to try and fix the problem.

1. Changed several times the thermostatic expansion valve
2. Cleaned the evaporator on a monthly basis
3. Changed the air filters on a weekly basis
4. Changed the fan's motor (direct action) for one with higher speed
5. Changed the evaporator fan's flywheel and motor pulley (indirect action) to increase the airflow through it
6. Etc.

Without obtaining any positive result.

Thus, he sent me directly to the job with these words: "See what you can find."

This problem (at the time) was probably one of the most interesting works I have done. It was, as I said, in a computer room located in a building, which was the base of an association called CARE, on the corner of First Avenue and Thirty-Second Street.

As usual, on my way to the job I was thinking of reasons that may cause this "frozen evaporator problem." I arrived at the place at about 8:00 a.m.

At the information desk (first floor), they informed me that the computer room was located on the sixth floor. I took the elevator and went to the mentioned floor; I entered the room. The first thing I saw at one corner of the space was a big chunk of ice hanging from the ceiling. And as I did not see anybody, I asked aloud, "Anybody here?"

A man's head appeared someplace in the room. He looked at me (he saw the company's emblem on the left side of my chest on my uniform) and without saying a word, with his forefinger indicated toward the flat sky (the roof). I understood that the A/C equipment was there.

I went upstairs. In the middle of the roof, just above the computer room, was a five-ton air-cooled condensing A/C. Unit with a sealed type compressor. Usually this size of unit is equipped with a sight-glass to observe oil level.

To gain access to the compressor unit, I removed one of the machine's covers. I was able to observe that the entire unit, as well as the suction line, covered with ice.

As per the equipment "information,"

Questions

1. Could you list the reason or reasons for a problem like this to exist for so long?

2. How many reasons could you list for this "frozen evaporator" problem in A/C applications?

3. What should be the first step in our troubleshooting process?

4. Besides what we see, are there any questions that we may ask of the person in charge?

5. How is it possible that a problem like this had passed unsolved by the entire team of mechanics in this company?

6. In our investigation, what should be the part or condition that we must observe that could probably help us in solving this problem?

7. What other questions may we ask to ourselves to shed light on solving this problem?

We can continue asking questions indefinitely, but let's get to the core of this problem.

===

EXPERIENCE 34-A

In the summer of the following year, and at this same place (Experience 33), the temperature was about 82 °F. I was sent to resolve the problem, which was affecting the entire building. This was basically a "noncooling" services call.

At the dispatcher's suggestion, I went to the "engineer's" office. When I arrived, he was seated at a writing desk, his legs on top of it.

He gave me a list of several systems on different floors and in various locations throughout the building where the A/C units were not cooling well.

He suggested I clean the condensers. He told me that many of them had already been cleaned in the last several days, in at least six visits of the same number of mechanics.

As usual, I wanted to conduct my own investigation, so I went to two of the mentioned floors in the list, and I touched the water supply and return lines in a couple of the A/C systems. I found them hot and almost at the same temperature.

With the information given by the various water supply/return temperature lines in the various A/C systems,

Questions

1. If you were the technician attending this service call,

   a) What should be your diagnostic?
   b) What should you do?

2. When using water from a cooling tower, what it is supposed to be the temperature difference between the water supply and return lines?

3. Should we verify the other equipment on other floors?

4. What should be our next move in solving this complaint?

5. Do we have any other questions for the "engineer"?

6. According to New York City's geographical location, in summertime, what should be the lower water temperature obtained from the cooling tower?

===

EXPERIENCE 35-A

This service call was for a meat market. In this place, there were two walk-in refrigerating rooms, one room (#1) for low-temperature (-15°F) meat products storage, the other room (#2) for medium-temperature (38°F) for processing (cutting, packing) meat products.

As they explained to me, the frozen meat from (room #1) was placed in the second room a day earlier.

The following day, in the same space (room #2), the meat products were processed (cut, packed) into smaller portions for sale.

The problem was that the second room, just like the first one, was freezing the product.

This equipment had the following characteristics:

1. Both rooms were served by the same refrigeration system, using refrigerant 502.
2. A semi-hermetic 5-ton compressor was used.
3. Power supply for the equipment was 220 volts, single phase, 60 Hertz.
4. An inside cooling tower supplied water to a shell-tube condenser type at 80°F.
5. The two rooms were equipped with solenoid valves in their liquid lines.

   Thermostats in each room controlled their operation.

6. The system was equipped with a dual high-/low-pressure control.
7. Room number two had an EPR at the evaporator's outlet.

Questions

1. In a system like this, what device and where is located to separate the room temperatures?

2. The low-pressure control's settings must be adjusted according to what room operating pressure?

3. What should be the boiling point for the refrigerant in the 38°F storeroom?

4. How do you verify the operating conditions in room #2?

5. What should be the operating pressure characteristics of this system?

===

EXPERIENCE 36-A

This service call was to a restaurant located in the village section of New York. It was near one o'clock in the afternoon. The place was empty. The equipment (compressor and evaporator fan motor) was running.

According to the person in charge, for the last two days, although the system was working, it was not cooling. The customer also complained of a current "passing effect" when the system's frame was touched.

System characteristics:

1. It was a 5-ton seal unit A/C system equipped with a remote air-cooled condenser.
2. The compressor was located underneath the evaporator's dripping tray.
3. Both motors, compressor and evaporator fan, had the same electrical characteristics: 230 volts, single phase, 60 Hz.
4. The system's ID plate said R-22
5. It was equipped with a dual-pressure control.
6. The compressor was equipped with a resistance around the crankcase's perimeter.

The equipment was de-energized, and the manifold was installed. The system did not have any refrigerant.

The problem was analyzed; the system had an electric current passage when touching it.

The system was pressurized (five pounds R-22) and tested for leaks. A large refrigerant leak was found quickly in the suction line, at the point where the suction/liquid lines passed through the evaporator's dripping tray.

Vibration wore out the rubber insulation covering the suction line, and caused damage to the tube. The unit's wiring was verified, and the obtained resistance readings were normal, although the ohmmeter showed a partial "ground" reading.

Questions

1. If equipped with a dual pressure control and there is no refrigerant in the system, why was the compressor running?

2. Why was there a "ground" reading?

3. What should we do next?

4. What recommended procedure should be performed?

5. Other than fixing the refrigerant leak, do you think there is any other procedure to be performed here?

Too many questions, right? Let's get the answers to these questions.

EXPERIENCE 37-A

Between the autumn and the spring seasons, work (air conditioning) is a little scarce. We perform routine or maintenance services, usually at the hours that the dispatcher believes advisable, not always for the clients' convenience.

This time, the maintenance was for an Italian restaurant. I arrived at the place at about 11:00 in the morning. There was no activity in the restaurant. I started my work by removing the system's covers.

System characteristics:

1. This was an A/C air-cooled condenser system of 5 tons.
2. It had a semi-hermetic unit.
3. It was provided with two air filters.
4. It worked with 230 volts, single phase, and 60 Hz.

The usual items to be accomplished were:

a) Clean the evaporator with water, brush, etc.
b) Clean the air-cooled condenser with water, some type of detergent, etc.
c) Change filters.
d) Check the fan belts.
e) Verify the pressure control adjustment.
f) Lubrication of bearings.
g) Etc.

I was doing that for about an hour when some customers began to arrive. Until this moment, I had no problems.

All of a sudden this man, a busboy, manager, or whatever he was, came to me, and in a bossy way, demanded that I finish working because I was interfering with his work.

I finished as fast as I could, left the job incomplete, but not before he had called to the office to complain about my work.

He made me go to the telephone, and the dispatcher asked me to try to finish fast. Thus, I did and I left the place.

Questions

1. What should be your reaction to this situation?

2. To prevent a situation like this, what should you do the next time?

===

EXPERIENCE 38-A

It was the month of May. By this time, the most important work consisted of preparing the equipment for the approaching summer. This time, I was sent to a pharmacy in the vicinity of Times Square and Seventh Avenue.

System characteristics:

1. The semi-hermetic unit had a 15-ton capacity and was located in an attic.
2. At the backyard was a cooling tower.
3. The evaporator was located in the same location as the unit.
4. The electric characteristics were as follows:

   a) The 15-ton unit used 440 volts, three phases, 60 Hz.
   b) The evaporator fan motor 230 volts, single phase, 60 Hz.
   c) Cooling tower fan motor 230 volts, single phase, 60 Hz.
   d) Water pump, located aside the cooling tower 440 volts, 5 HP, three phases, 60 Hz.

The job included cleaning and filling the cooling tower, bleeding the circulating water pump lines, lubricating all motor bearings, verifying/adjusting belts, fitting controls, etc.

As usual, I did not waste any time, and although I was busy performing my job, I noticed that the man in charge (possibly the owner) kept roaming by where I was working.

When the work was almost finished, a company supervisor arrived. He took me aside and told me that the client had called the office and complained that when I arrived I did not bring any toolbox. I started to explain to him my point of view, but he said to me, "Jose, I know how you work. I agree with you, but the client wants to see a box of tools." He suggested to me that next time I bring an empty box with me.

Meanwhile, the "client" smiled, seeing that he had caused me to be reprimanded or had my attention called.

I thought that was OK. I did not want to cause any bad blood.

==

EXPERIENCE 39-A

In the year 1970, this time I was sent (at 10:00 in the morning) to the Harvard Club located at Forty-Sixth Street, between Madison and Sixth avenues. They have been without A/C for several days; at my arrival, the person in charge told me that since the previous mechanic changed a control, the equipment had not been performing well.

When walking into the machine room, I notice that in the upper part of the compressor a new dual (discharge, low) pressure control was installed. I did not know why, but the first thing I went to verify was the mechanical (tubing) control's connections.

I noticed immediately the wrong connection: the low-pressure control's tubing was connected to the system's high-pressure side, while the high-pressure control's tubing was connected to the system's low-pressure side.

System characteristics:

1. The compressor was located on top of the water-cooled condenser/receiver (sized about 6 feet long by 3 1/2 or 4 feet in diameter)
2. The reciprocant compressor was a W open type.
3. The system capacity was 200 tons. Therefore, the refrigerant charge was of approximately a minimum of 900 pounds of R-12.
4. The cooling tower I supposed was in an outside location. Connected to the condenser's water outlet was a water-regulating valve (WRV).
5. The system's compressor and condenser were equipped with service packing gland type valves.

6. Here was how the dual pressure control was supposed to be connected:

   a) The LP control to a T adapted on the suction service packing gland valve type.
   b) The HP control connected directly to the discharge line side.

To connect this control (for safety reasons), it is recommended not to use any valves between the system and the control.

Since we already noticed the wrong connection of the dual-pressure control,

Questions

   1. What should we do to repair this "mistake"?

   2. Is there any special procedure to be performed to exchange the HP control connections?

   3. What should we suggest doing?

   4. Is there any easy way to perform this repair?

   5. When calling the office, besides reporting the problem, what should we ask for?

   6. To perform a job like this, how much time may we spend?

==

EXPERIENCE 40-A

It was the month of July. The work was at a fish market on Second Avenue and Seventy-Second Street.

The problem: A crushed ice machine, which produced the ice used to maintain the lobsters, clams, oysters, and other seafood products in the display counter, was not working.

According to the customer, this was the third time this year that the machine showed the same problem (the last incident was about a month ago), rupture of auger, a device that "scrapes" the formed ice from the internal evaporator's surface.

In this ice-making machine, the auger, approximately 2 1/2 inches in diameter, was found broken into two pieces, as if it were made of cardboard or something similar. This device does the job through a gear transmission that reduces the speed from the electric motor's 1,750 rpm to approximately 6 or 7 rpm for the auger.

At my request, and according to the client, this equipment had these other characteristics:

1. It produced approximately 500 kg (1,100 lb) of ice in twenty-four hours.
2. The power supply of 230 volts, single phase, and 60 Hz.
3. The equipment used R-12.
4. Used city water (temp. approximately 60°F) in a condenser shell and coil type.

I called the office to report the result of this service call. They already knew the Information required to order the new auger.

Questions

1. What do you think is the reason for this problem, which happened three times in six months?

2. Should we worry about finding an answer for this problem?

3. Are there are any other questions that we can ask?

4. In normal operating conditions, what are the operating pressures supposed to be?

5. What should be the refrigerant's boiling temperature for an application like this?

EXPERIENCE 41-A

It was summertime. The temperature at this time (about 11:30 a.m.) was already 78°F. I had already been in this place, an Italian restaurant, last spring to perform a maintenance service on the A/C equipment. I remembered clearly, as well, that here I almost got removed or "kicked out." I had to work fast, and left the maintenance service incomplete. Today, I met the same guy that treated me like a "dog" at that time.

Now he was completely changed, I may say different. He was almost a supplicant, I would say. He needed my services. The equipment was not working properly. It was *not cooling enough*, and he confided in me that for the last three days the customers have diminished.

As we may remember, this A/C system used R-22.

Just for the record, in a service call like this,

Questions

1. What questions may we formulate for this person?

2. What should be the points that we concentrate our attention on to find the problem?

3. Do you think it is a very good idea to have or to remember the history of every job?

===

EXPERIENCE 42-A

For this service call, one of the supervisors wanted to talk to me. According to him, the problem was that somehow water from the cooling tower (located on the second floor) was dripping to the first-floor store, causing damage to the stored merchandise.

He also told me that where the cooling tower was, several inches of water had been pooling for a long time. They had had to replace and fix the floor, as well as change the cooling tower's unloading air duct work several times because it got oxidized and severely damaged.

I do not understand certain procedures and problems I found in the performance of my job here. I never wanted to express any criticism, but some of the work I had to do were for problems that had been there for a time, longer than anybody can explain.

One becomes aware of interesting things, and in this case, it was new for me. In this fashionable warehouse located in Fifth Avenue, I learned that the rent, considering the size of the area, its location, and that this was in the early seventies, was pretty low, a rate of $150 per square foot.

In order to provide the place with A/C service, they had to install the cooling tower inside the building, as mentioned, on the second floor. It was installed about four feet away from the window that opened out to the street. The cold fresh air entered the cooling tower through an open window; and to allow the unloading of the air back to the street, a metal ductwork was used from the top of the tower to another window.

Questions

1. What do you think may be the reason for the problem expressed here?

2. Why did it take so long to find the cause of a problem like this?

3. What should we do to resolve such a complaint?

4. Besides the information given to us, are there any other questions that can be asked?

==

EXPERIENCE 43-A

This morning, the dispatcher sent me to the company's warehouse to take a condensing unit for a beverage cooler device in use at an Italian restaurant. According to the dispatcher, they had been waiting for this equipment for three weeks.

The equipment was taken from the store and carried out to the truck, then I drove to the restaurant to do the repair. When I got there, I walked in, and the person in charge asked me if I had brought the equipment. I answered him yes and told him that I had it in the truck. He was very happy because they had many problems caused by the lack of that cooler device.

I wanted to know where the equipment was and where that unit was to be changed. He called an employee and asked him to take me to the cooler's location. I followed the employee to a lower floor. There, against the wall of a large and wide corridor, I saw the beverage cooler device.

Thought: I always knew and I know now that we are thinking humans, not robots. And whenever we are ordered to do something, we have the right, and we must exercise it, to verify what we are told before we begin a job.

Since I have encountered so many jobs in which the problems keep reoccurring for so long, every time I was sent to perform a service it seemed to me a very good idea to perform the job in the most complete and perfect way possible, without limiting myself to just doing whatever we were ordered to do.

Questions

1. Would you change the condensing unit right away as you were told to do?

2. May we conduct our own investigation of the problem by ourselves?

   If yes, what should we do?

3. What procedure should we perform next to verify this job?

4. Are you curious about the condition of the old system? I was.

===

EXPERIENCE 44-A

We are at the height of summer; the temperatures are high today (92°F). The service call was to the same pharmacy on Times Square (two months have passed since I was there) where I met my "my smiling friend" (Experience 38).

This time, I had an empty toolbox in my hand.

The customer greeted me this time with a worried and not-so-big smile. He begged me to "please" do something, because this place (according to him) was hell! He said the equipment was working wonderfully until yesterday. Since then, it hasn't been working.

I thought, "Do people have a bad memory, or what," as they change their mind for no reason. One time they treat you like trash, and the next (like in this case) they'll even try flattery.

Just a reminder, the electrical characteristics were:

a) A semi-hermetic 15-ton unit using 440 volts, three phases, 60 Hz.
b) The evaporator fan motor 230 volts, single phase, 60 Hz.
c) Cooling tower fan motor 230 volts, three phases, 60 Hz.
d) Water pump, located aside the cooling tower 440 volts, 5 HP, three phases, 60 Hz.

Questions

1. Do you remember the customer with his "smiling" behavior from last time?

2. What should we do now?

3. What kind of tests do we perform?

==

EXPERIENCE 45-A

This was not the first time I have gone to this club (Friars), but this time I am to repair an A/C system by changing a broken compressor. The system was located at the back of the club, and the compressor was mounted on a metallic platform at a height of about 30 inches off the floor. This compressor weighed approximately 250 pounds. I called the office to send somebody to help me raise the compressor.

This system had the following characteristics:

1. A 5-ton compressor of a semi-hermetic type.
2. It was connected to an electrical circuit: 440 volts, three phases, and 60 Hz.
3. It used R-22.

In the meantime, while waiting for help, I verified the electrical part and it was OK. I installed the manifold gauges and ran the unit. It was not pumping (mechanical failure), so it had to be changed.

I isolated the compressor from the system by front-seating both suction and discharge packing gland service valves, so it was ready for the dismount and change.

Time went by, about two hours, and nobody showed up.

Questions

1. What should we do, wait or what? Too many questions, right?

2. Should we try to raise the compressor by ourselves?

3. Should we call the office again, or is that a waste of time?

===

EXPERIENCE 46-A

The dispatcher told me to be on standby so that when an ordered compressor arrived I could go replace the one at a bank located at 125th Street and Park Avenue.

About three weeks passed, and then one morning, at 7:45, I called in find out where I had to go that day. The dispatcher asked me if I remembered what he told me about the compressor that had to be changed in the bank. To be honest, I didn't.

Anyway, he told me that Brown, a helper, had the compressor in his truck and he was waiting for me at the bank. The dispatcher gave me the address. When I arrived there, Brown was standing at a street corner, and he indicated where I could park my truck.

I followed where he indicated. Then he approached my truck's window, greeted me, and asked, "OK, Jose, what are we going to do first? Are we going to have some coffee? Or are we going to bring down that baby to its workplace?"

I answered, "Neither one nor the other." He looked at me in a strange way. Then he asked me, "Why?" I answered, "Because I want to see the compressor's condition myself."

Then, in dismay, he told me, "Oh, come on, Jose!" He enumerated to me the names of six different mechanics and a supervisor who had been here, and he said all of them agreed that the unit is burned out. I responded to him that I want to verify this myself anyway.

I took from the truck my tools: a pair of pliers, screwdriver, lantern, and an electric multi-meter, and I was on my way. He followed me a little reluctantly.

It being so early, the bank was still closed. We knocked at the front door. A guard opened the door and we got in. I saw the floor. There were three wide marble steps leading down to the tellers' windows.

The person who opened the door indicated the location of the equipment room. When entering the place, I observed on the floor two similar compressors located side by side. The one on the left was working.

Comment: Let me make it clear that I have never set out to be contradictory or a troublemaker. To avoid being misjudged, on a few occasions I did whatever I was told. But having seen so many "mistakes," I sometimes decide to do things my way. This was one of those cases.

A/C system characteristics:

1. The compressor that we were going to change was one of a set of two 15-ton semi-hermetic devices. They both had air-cooled condensers at the window of the same room.
2. These units had the same electrical characteristics:

   a) 440 volts   b) Three phases   c) 60 Hertz

3. The refrigerant used was R-22.
4. Each of the two electric circuits had separate three-phase contactors, as well as circuit breakers with the same characteristics.

Since we had the replacement compressor and all the necessary parts/components, and according to Brown several mechanics and a supervisor had diagnosed this as a *burned-out motor*,

Questions

1. What should we do? Change the compressor without "wasting" time?

2. Or should we conduct our own investigation?

   If yes,

3. What should we do next?

4. What procedures should we perform?

5. What tools should we use to conduct our investigation?

==

EXPERIENCE 47-A

For a week now the company has assigned me to work at night; my duty started at 4:00 p.m. and finished at midnight. It was Saturday. At about 6:30 p.m., I called the service operator, and she provided me my first service call. It was to an Italian restaurant. The customer had complained that the A/C system was not cooling. At this hour, the restaurant was getting busy, and one of the air-conditioned spaces (according to the client, there were at least two of them) was a mezzanine. They indicated that the equipment's condensing unit was located next to the restaurant's kitchen (quite distant from the mezzanine). I got there, and the place was hot. When I was passing through the kitchen, I saw a man preparing some cookies (crackers) or candy on trays next to a working stove.

I found the compressor. The evaporator was at the same location. I also found the cause of the problem: a refrigerant leak (oil trace) in a flare-nut connection in the expansion valve inlet. I got the appropriate wrenches and attempted to tighten it, but this did not have any effect, and the flaring connection had to be remade.

System characteristics:

1. It was an old A/C system using R-12; I couldn't find its capacity, but by the size of the set, an open-type compressor mounted on top of the water-cooled condenser/receiver, I assumed the equipment had a refrigerant charge of no less than 200 pounds.

2. The combination water-cooled condenser-receiver was about four feet long and approximately two feet in diameter; it had a cut-off valve at the condenser's inlet, as well as a service (king) valve at its outlet.

3. The condenser was equipped with a water-regulating valve (WRV) at its outlet.

4. The compressor's motor had these characteristics:

   a) 230 volts   b) Three phases   c) 60 Hz

5. The thermostatic expansion valve where the problem was found had a capacity of 15 tons.

6. As mentioned above, this system had at least two evaporators.

With the information above,

Questions

1. What procedure do you think has to be performed to fix this leak?

2. Do you know what is the name given to the valve located at the condenser's inlet and what is the purpose of that valve there?

3. According to the client, the mezzanine was distant from the condensing unit; the other evaporator was the one located beside the unit. Do you think the condenser/receiver set was big enough to store the entire system's refrigerant charge?

4. What procedure or procedures may we recommend performing?

5. What precautions may we observe here?

===

EXPERIENCE 48-A

I was still working the nights/weekend service; it was 8:00 a.m. on Sunday. I called the service operator, and she gave me the only service call she had. It was in a restaurant in Massapequa, somewhere on Long Island. As I was new to this city, I did not have the slightest idea how long or how big Long Island was. I accepted the call. I started to drive.

By following a Long Island map, at about 11:30 in the morning I finally arrived at the restaurant. I entered the place and asked for the manager. He thanked God that I had arrived, took me to the kitchen, and there showed me a cooler that according to him had not worked since morning of the previous day. I reviewed the equipment and found that the only problem was that somebody had disconnected it from the power supply outlet.

Thus, I let the manager know.

He said, "I'm very sorry." Then he asked me if I knew anything about heating, because the previous night they had to suspend a party because the place was too cold. I answered him yes, but he had to sign a service order for that.

He said there was no problem with that; he would sign anything.

Questions

1. What should we do to fix this problem?

2. Should we have any other questions for the client?

3. What tools do we need to perform this job?

==

EXPERIENCE 49-A

During all these years, I have kept the same home telephone number. This way, my former students can always communicate with me. Sometimes, they have problems with their works and call me to consult what they can do. Until this year, I have received (3) calls from Argentina, (6) from Dominican Republic, (several more) from Puerto Rico; no counting the ones from the City or USA territory.

Although, the most peculiar of them have been from a Panamanian former student, whom after he finished the course, in my school; he went away with his family to California. Once there, he registered himself in a college to continue his studies in this trade. In the first call, he told me that he had some problems with some of the professors who, according to him, taught some topics different, from as he had learned with me.

I tried to clarify those topics to him on the best possible form; he was satisfied. After the first one, four more calls took place (approximately one per week).

One day he called and put me in contact with one of his college professors. This man an (American) told me that by some time he had the idea to speak with me; because of this student gave him some "Hard-time" and that he had woken up his curiosity and it wanted to speak with me. He told me that the Panamanian one was a "tremendous" student, very understood and that he professed an admiration for me without limits.

"This professor consulted me" several things and I discussed with him, "to dissipate" some doubts that he had. This call lasted more than 2 hours.

The professor invited me, if sometime I went to California so that we were able to share ideas.

There are no comments or Part "B" for this experience. I think it is self-explanatory.

Anyway, I continue having the same telephone number. Had been a while that I do not get calls from my former students.

===

EXPERIENCE 50-A

We were at the beginning of spring. On this day I was sent to a fast-food restaurant. According to the dispatcher, this problem had existed for longer than a week.

The place was located at the front of the bus terminal in Manhattan, Eighth Avenue and Forty-First Street. I entered the place. At this hour, 8:30 a.m., it was empty. The employees were seated and had nothing to do. The temperature in the place (I looked at the thermostat on the wall) was at 100°F. I asked for the person in charge. A Hispanic man appeared and sarcastically said to me, "And you think that you can do something? Even if the **other mechanics** could not?" It bothered me, the contemptuous way in which he greeted me, and I answered, "You simply tell me the problem and limit yourself to showing me where the —— equipment is located, and I will show you what I should be able to do."

I followed him to the back of the store. Once there, he pointed at the stopped A/C system. It was located in the center of a 40-foot-by-40-foot space. This person informed me that two mechanics from the company responding to the same complaint had been working here for a couple of days and making changes to the heating system. No matter what they did, the high temperature in the space remained the same.

A/C equipment characteristics:

1. This was a 7.5-ton air-cooled system.
2. The seal unit had these electrical features: 230 volts, three phases, and 60 Hz.
3. On the left side of the A/C equipment were two steam supply and return cast-iron pipes to provide heating when required. I

93

noticed as well that portions of the two-inch piping, as well as the steam valve control, had been changed recently.

I turned the A/C fan motor on. I placed my thermometer first in the A/C air intake, then in the air outlet ductwork to the space. The temperature readings were the same: +/- 100 ºF in both cases.

According to the information supplied,

Questions

1. What can be producing the high temperatures in this place?

2. Besides the information received, what should we ask next?

3. What test should we perform next?

4. Other than cooling, heating, cleaning, air movement/ distribution, etc., what other features should an A/C system have?

5. What other condition should we verify?

==

EXPERIENCE 51-A

It was Christmas time, and I was sent with another mechanic (to help me). He had been in this place before, a French restaurant, and he knew what the problem was. We arrived at about ten o'clock in the morning. The heating equipment, supplying heat to two spaces, the reception and the restaurant's dining area was not working well.

The owner seemed annoying enough, but he explained to me the problem. For several days they had experienced problems with the heating system; the equipment was located in the ceiling, above the reception's area. He said that several service people had been here, trying to fix the problem, but no result.

I never saw a kitchen and a dining room so clean and pleasant. The owner himself prepared the place for lunchtime. He put fresh flowers on each table, and was aware of the minimum detail.

In the heating equipment, somebody had been struggling with the electrical installation and confused everything. It was a mess.

With the system's electrical printing diagram and the knowledge the mechanic had of this equipment, we were able to have a better idea of what we had to do.

With the information supplied, by the customer . . .

Questions

1. What else should we had to ask?

2. What procedure should we use to fix this complaint?

3. What tools may we use, to put this equipment back in operation?

===

EXPERIENCE 52-A

To this restaurant, some mechanic came every day. It seems that they like to come here, because the Chinese owner, a very friendly one, offered drinks to each of them every time they come. He did the same with me, but I did not have any desire to drink. The place felt well, but the client said; there was a problem: He said that this is happening for quite some time.

"The thermostat on the wall, never registered the correct reading" (He had its own thermometer). The thermostat, according to him, it always gave a higher reading.

This complaint seems quite simple; all we had to do is to compare the wall's thermostat reading with our own thermometer. And make sure, they show the same reading.

According to the "Book," the customer is always right; let's see what we can do.

Questions

1. What would you do to satisfy the customer?

2. Was the thermostat at the wall giving the right reading?

3. What procedure do we perform to fix this complaint?

==

EXPERIENCE 53-A

Many times, it is difficult to understand people. In my school, we had a Colombian student on his twenties; his sister was a resident here; she requested him before the immigration office and as it is a law, she had to be responsible for his staying.

The problem was that he pretended the sister, who was already paying its tuition, to provide absolutely everything to him. I was trying to explain him, that he had to cooperate, but he continued with its attitude. I decided not to take sides in this matter anymore.

The school had several low capacities air conditioner systems for the training. One day, the youngster requested to me as a favor, to exchange an A/C a little smaller for one that he had in his house. I did not see anything wrong, and I acceded. He brought its equipment of 1.5 tons in very good conditions, and he took one of ¾ of a ton; that worked well and it was in very good shape as well.

A couple of weeks later, it entered the school a man looking for an A/C system for his business; a beauty parlor, located in Amsterdam Avenue at just two blocks away from the school.

The customer supplied the following information: size of the space, number of occupants, total heat load, available power supply, etc. We thought the A/C system of one and a half tons capacity that the student had brought should be able to make the job. We arranged the price including the installation; I let the students know that the entire product of that transaction should be saved, for the celebration that would usually occur at the end of each course. All were in agreement with that project.

One afternoon, at about 6:00 p.m., we installed the equipment and we left it working well. However, on the following day, the client came to complain; the equipment did not work.

I told him that we had to wait until the students arrive so that they can help me; he acceded.

At about six o'clock in the afternoon, with a couple of students, we went to the place and we put the equipment to work. It worked well. I said to the client if he noticed something wrong; let it me know immediately.

About one hour had passed and the client came back. With several students, we went to the place and found the equipment was cycling, by the thermal overload; we verified the running amperage and the operating pressures. They were high.

I know, for you it may not be enough information, but let's give it a try.

Questions

1. What do you think may be the problem here? Think in as many causes that you can.

2. What test do you suggest to be performed, to find the problem?

3. Do you think we may take the equipment back to the school, to verify its operation?

Who knows? Probably the problem had nothing to do with the equipment. Let's see.

===

EXPERIENCE 54-A

According to the dispatcher, several mechanics (in a time of about 3 weeks) have been sent to fix an A/C equipment for this computer room; but, no one had been able to find "a possible" refrigerant leak.

The system was an old one, having the following characteristics:

1. The electrical features were: 440 volts, 60 Hz, 3 Ph.
2. It worked with R-12
3. Had a hermetic type compressor of about 10 tons.
4. It uses an air-cooled condenser, located in the same room, against a window.
5. Below the condenser a receiver, about 3 1/2' long by 2 feet diameter, equipped at its outlet with a service packing gland type (king) valve.
6. System equipped, with a dual pressure control type.

At my arrival, the equipment was not working, but I notice (by hand touching) that the compressor was lukewarm as if just had stopped. I backed up the seat of the small packing gland valve, connected directly to the compressor's body; in which the low-pressure control was connected. Proceeded to disconnect the low-pressure control, and in the same place by using a "Tee" adapter, the manifold's hose of the low-pressure gauge, as well as the low pressure control were connected. The valve, was opened again, the gauge gave me a low suction pressure reading (about 25 psig).

It was observed as well, the suction line was attached to the compressor body; by a flange and four bolts and it had no service connections. As it was a service (king) valve, at the condenser-receiver's outlet; I make sure the valve was in its back-seated position. Then, the manifold's hose of the high-pressure side was connected to that valve, when opened; the pressure was normal, according to the ambient temperature (84°F.) in the place.

Within a few minutes I noticed that the suction pressure increased although very slowly.

Through the manifold, from the system's high to low pressure sides a little refrigerant was passed, to activate the low-pressure control; and the equipment started; it worked for a few seconds and stopped again, by the action of the low-pressure control.

Waited for about 15 minutes and the suction pressure went up, activated the low-pressure control; the equipment started to work again, by a few seconds; then, stopped.

Questions

1. What do you think may be the reasons for the "cycling" of this compressor? Can do you think, in at least two of them?

2. What should be our next "move"?

3. What other procedures should we use to find out the problem?

4. What should be the LP switch cut-in and cut-out set points?

==

EXPERIENCE 55-A

This time, I just finished a service call in a restaurant located at Madison Avenue and 41st. Street. There, I repaired and verified the operation of the A/C system.

That equipment had the following characteristics:

1. It was a 5-ton system with an air-cooled condenser.
2. Electrical characteristics 230 volts, 60 Hz, three phases.
3. It used R-22.
4. The ambient temperature was about 80°F.

Through the manifold pressure gauges readings; the unit shows a little lack of refrigerant. By using an electronic leak detector, the system was checked for leaks; found one at the thermostatic expansion valve flare-nut outlet. That connection was retightened and the leak was fixed. Refrigerant was added, and the entire system's operation was verified.

When that service call was finished, I got in touch with the dispatcher who sent me to another place only half of a block away from where I was.

This was a restaurant as well, and the service call was similar to the one that I just left.

"Not cooling enough." I checked the operation of the A/C. system. This equipment was in appearance, very similar to the one I just served. The manifold was installed, put the system into operation and in the course of the inspection, by using the same procedure as in the last service call; I found a refrigerant leak in one of the flare nuts, this time in the filter-dryer. I tighten the connection and repair the leak. The manifold gauge operating pressure readings, showed what I "assumed," a low suction pressure reading (+/-40 psig); indicating a "lack" of refrigerant.

By my negligence, without making myself sure, I assumed that this equipment used the same R-22 as the one A/C that I just serviced. I had just added a little of such refrigerant; when I raise my eyes to see the evaporator, too late, the color in the TXV's sticker; it was of a "Yellow" color = ID color for R-500. Not popular at this time, in A/C systems.

Note: Comparison: For a boiling point of 40°F (A/C applications):

R-22 has a pressure of 68.6 psig
R-500 has one equal to 48.2 psig

Questions

1. What should we do to prevent mistakes like this one?

2. Now, what procedure or procedures should we perform?

3. What do you think were the important tools used here?

4. What should be the operating discharge pressure for this refrigerant?

5. Since we do not have the information to calculate the charge of refrigerant, what should be the other reference points (suction and liquid line temperature) values?

6. Which should be, the different ways to identify the type of refrigerant in a system?

==

EXPERIENCE 56-A

There were about 12:30 in the afternoon; I had just finished a service call, in a restaurant located at Madison Avenue and 41st Street. I called the office and the dispatcher told me, that one of the supervisors wanted to speak with me.

The dispatcher put him to the phone and he told me:

"Jose, take all your time, have lunch, and later go to the computer room in the second floor of ### in the 6th avenue. There is a mechanic working in the A/C! I'll meet you there."

I said OK.

In those days, and with the permission of the company, I had working with me (to gain some practical experience) one of my students.

Yes. I took all my time; furthermore, between the lunch and getting a place to park the truck, I arrived at the computer's room almost at two o'clock afternoon. The supervisor was already there.

According to his appearance, he had been there for a while; he was already sweating, I thought that he had 30 minutes or more that he had arrived; he turned to see me, I noticed him a little annoying but he didn't say a word.

On the floor, there was the A/C system's covers, tools, and all kind of materials.

I approached the mechanic, and I asked him what the problem was. He told me that "We do not know why, but the system does not start."

Within few minutes, I got the information of the equipment: This was a 15 tons system.

1. 440 volts, three phases, and 60 Hz.
2. Control circuit of 24 volts.
3. It had a Tube within a Tube type, water-cooled condenser.
4. Used R-22.
5. As far as I saw, the system used several types of electrical controls: thermal overloads, high- and low-pressure controls, a lockout relay, water-flow switch, low-limit sensor, etc.

Questions

1. What should we do?

2. What procedure we should perform, to find the answer of the system "does not start"?

3. What type of tools should we use to troubleshoot this equipment, now?

4. What specific precautions we must exercise when using the tools?

5. If we were using any electrical instruments, what are the electric rules to be followed?

==

EXPERIENCE 57-A

The service call was to a computer's room in 6th. Avenue, in a building bordering the Radio City Music Hall. Armed with my usual tools, entered the place. I met the A/C person in charge, he was infuriated. He complained that one week has passed when a mechanic of the company made some repair; since then, the compressor "runs, but the equipment does not cool."

He informed me as well that several mechanics have made different types of works in the equipment:

Up to what he understood, he enumerated:

1. Removed the refrigerant.
2. Pulled vacuum to the system.
3. Charge refrigerant again. These two procedures (2 and 3) have been repeated several times.
4. Cleaned the condenser (upstairs)
5. Others.

But the equipment continues the same; it said to me that part of the equipment (the condenser) was located on the second floor and, to arrive at it, I had to go through an office; he gave me the number of it.

The system on the first floor had the following features:

1. It was a split system with a semi hermetic compressor with a capacity of 15 tons.
2. The electrical characteristics were: 440 volts, 3 phases, and 60 Hz.
3. It used R-22.

The compressor was in a space beside the computers room. I took the manifold and installed it to the compressor's service connecting valves. I checked the operating pressures; both of them were very high.

I looked for the "green" cards used to log the service events performed to the systems. I couldn't see them, and I felt uncomfortable to ask the person. He was angry and probably he didn't even know it.

I went to the second floor and knocked on the office's door, and a lady secretary, very upset, I might say angry, looking at the name in my uniform's chest, said, "You people again? When are you going to finish? It was already necessary to change the carpet once. You must remove your shoes to enter."

I politely answered her, "Young lady, this is my first time here. Nevertheless, I'll try that it is the last time you have to deal with us."

I removed the shoes and I went to a window, through which I could get into the place where the air-cooled condenser was located. Once in the other side of the window, I wear my shoes again.

The remote condenser here used two permanent split capacitor (PSC) type motors.

At this point, with the above information . . .

Questions

1. Why do you think, this system shows; those high operating pressures?

2. What type of work was performed by the first mechanic? And where? I couldn't find the answer by asking.

3. What should you do to find the reason for this problem?

==

EXPERIENCE 58-A

In this occasion, I was sent to the WPIX Radio station; located in the second floor of a building in Park Avenue and 36th. Street; where the A/C equipment, it was not working well. This system was installed in the center of a quite big room. There were a lot of shelves, where all the station programming, acetate discs were stored.

The A/C system was running; it was not cooling at all.

Equipment characteristics:

1. This A/C system was a 5-ton air-cooled condensing type.
2. The electrical features were 230 volts, 1 phase, and 60 Hz.
3. The system used R-22 as refrigerant.

According to the person who greeted me, "The equipment was not cooling because of a lack of refrigerant," He said that this was the "zillionth time" they had that problem. I checked the system; indeed the system, although running, wasn't cooling. I went back to my truck and got the appropriate tools for this type of repair.

1. Soldering acetylene B tank set and implements (solder, flux, sandpaper, etc.).
2. Manifold gauges.
3. Vacuum pump.
4. An R-22 refrigerant cylinder.
5. A leak detector of the electronic type.

I stopped the system, removed the equipment's covers, and by using the appropriate procedure, installed the pressure gauges, added some refrigerant to verify that the system had enough pressure to conduct the leak detection. By using the leak detector, I found a refrigerant leak in the suction tubing that ran vertical (together with the liquid line) through the left-back internal corner of the dripping tray, located below the evaporator.

This job was one of those in which a refrigerant leak was found easily.

Questions

1. Do you have any valid idea why it wasn't repaired the "zillionth times" the system was service before?

2. Do you think there is any obstacle in performing such job?

===

EXPERIENCE 59-A

When the repair was finished (Experience 58), the person in charge asked me if I could verify the operation of other A/C equipment; in which they had problems for a long, long time. That was since its installation. I told him, to call the office so that they give me the authorization. Thus, he did it and they authorize me to (check) and if possible, make the repair.

According to what the people in the radio station told me, two or three times per week the mechanics of the installing company; came to add/charge refrigerant to this system.

This A/C system had the following fixtures:

1. It was a 3 tons split system, where the air cooled condensing unit; it was located at about 40 feet of the space the A/C system it served.
2. The electrical features: 230 volts, single phase and 60 hertz.
3. It used refrigerant 22.

I verified the equipment, and very easily found that the entire refrigerant surfaces of the discharge, liquid and suction lines, although they were new; were covered with oil. All of them were porous (leaking).

As far as I knew, the copper tubing used in all R & A/C. applications, are made with a very fine grain and are submitted to precise processes; to prevent refrigerant leaks. With "No joints" and "No porosity." Other metals, aluminum, bronze, iron, steel, brass, even copper, etc. are used to make all types of piping jobs; but, they do not require such precise requirements, as the ones needed in the R & A/C. applications.

Questions

1. How come this type of wrong installation was performed? Economy?

2. What should we do to prevent this type of mistake or problem?

3. Could we think that a lack of knowledge can result in this problem?

===

EXPERIENCE 60-A

We were in the summer season, this service call is to a grocery store; located at Audubon Avenue and 183rd. Street in Manhattan's upper side. The faulty equipment was a freezer, that according to the client it is not cooling; as a result, some of the stored product (meat) has been damaged.

They indicated me the location of the condensing unit, the cellar.

I went down a few stair steps, and I found myself, with a large amount of oil on the floor around the compressor.

This equipment had the following characteristics:

It was a device equipped with an air-cooled condenser.

1. The compressor, a semi hermetic type of 3/4 of a horsepower.
2. It uses Refrigerant 502.
3. The power source had this characteristic: 115 volts, single phase, and 60 cycles.

System's operation requirements:

The required temperature of these products was -10°F.

The temperature in this place (at this time) was near 98°F (36.7°C.).

I reviewed the compressor and found the source of the oil; it was due to the rupture of the oil-level sight glass in the compressor's crankcase (carter).

The operation of this system, under the present conditions, was in a very serious problem. Not only requires the repair, but, to do something to help the system to work under less stress.

Questions

1. Why do you think this may happen?

2. What are the operating pressures of this system?

3. When this system stops, what is the crankcase pressure at the temperature in this location?

4. What should we do/recommend to prevent this for happening again?

5. Do you have any idea of what to do, trying to remove such rough operating conditions from this system?

===

EXPERIENCE 61-A

By this time, I had one not very pleasing experience. I was sent to help a mechanic, to whom I met in Madison Avenue and 90th. Street.

When I arrived at the place, and met this individual, one of the old mechanics that I have seen around the shop; he told me that my aid was no longer necessary.

Without asking it, he confided me that the equipment that was going to be serviced, didn't have electrical energy, and that he had recommended the client to call the company, tell them that the service; was no longer required.

He said to me: that later at the end of the day, when he left the work he would return; fix the job and keep to him the money of this work.

To me, the behavior of this person wasn't honest at all, but I did not say anything because, I thought that should be my word against his; and me, who was new and did not dominate the language, could put myself into a problem.

In this world, anything could happen.

Questions

1. What do you think of this behavior?

2. What should be your reaction to this?

3. Could you discuss this behavior with someone in the company?

==

EXPERIENCE 62-A

This morning, at my arrival to the company's office; the dispatcher told me that one of the supervisors wants to talk to me; I went to his desk and he told me among others, several details related to this job:

1. In this place, a pharmacy, they had had problems for a long time (several years).

2. Finally one of the supervisors, convinced the customer to change of the A/C equipment.

3. The customer's only condition (because; the limited space) was that the new equipment have the same dimensions as the old one; so could be installed in the same location.

4. The supervisor told me as well that they had changed the A/C system, but for some reason, they couldn't get from the new equipment the proper cooling operation.

Then he asked me to go to the place to look and see if I could find the reason.

Before I left, I asked him for the system characteristics, which he recited to me:

1. This equipment had a 5-ton capacity self-contained system, water-cooled.
2. It worked with refrigerant 22.
3. It had a water-cooled condenser of the shell-and-coil type.
4. The compressor was of the sealed type.
5. The electrical features were: 230 volts, three phases, and 60 Hz.

I went to the pharmacy, located on 57th Street between 2nd and 3rd Avenues. I met the customer, and he was kind of annoyed (commented audibly for me "hum, one more"). He wondered what I was going to do. He complained, "You people are working for a week and already are interfering with my business."

What could I do? I just promised to him to do my best, to put an end to the problem. I asked him where the cooling tower was located; he told me that it was in a garage at 58th Street between 2nd and 3rd Avenues.

I went to the A/C equipment location; it was operating. I started to get familiar with it.

To verify the system's operation pressures readings, I installed the manifold gauges.

Note: This time, as I usually do in many other jobs, I performed this same procedure (as a disguise) to appear "busy," while I was thinking. So the customer does not notice that we are lost. This way, we can get familiar with the equipment characteristics and more.

When I "sensed" (by hand) the supply/return water temperatures piping at the condenser, I noted that the return one was warmer, but more important; there was very little temperature difference between them.

In normal operating conditions, this temperature difference (cooling tower) should be of approximately 10 to 12°F.

When taking the suction and discharge pressure readings, I noted they both were high.

Questions

1. Why do you think, they had that problem for so long?

   Note that the same problem occurred before, and is now occurring in the new equipment.

2. What do you think the problem is?

3. What procedure do you perform right here, to solve this "mystery"?

4. You see? Taking the operating pressure readings takes us nowhere near the solution. But it tells us something was wrong.

5. Where do you think may we get some more helping information?

==

EXPERIENCE 63-A

One of my former students, (Anthony, opened a domestic refrigeration shop. As I believe this information can be used in any of the R & A/C fields, I'll relate it here.

He bought household refrigerators (different capacities), in a scrap refrigeration metal place in New Jersey; and after repairing them, painting them, etc., put them to be seen by the people passing by the front of his shop. One of those afternoons, a man was interested in a two-door refrigerator.

According to Anthony's information, the refrigerator had the following features:

1.   A sealed unit of 1/3 hp.
2.   115 volts, 60 hertz, FLA 3.2 amperes.

The customer spoke to Anthony and they fixed the price, which included the transportation (two men would carry the refrigerator to the fifth floor of a building). (It would be worth the trouble to say that at this time in New York City the buildings this high did not have elevator service. This service was obligatory when the building had above this height). The following day, at about 6:00 p.m., they delivered the refrigerator.

Arriving at the apartment, they plugged in the device, which started and worked without problems. But on the following morning at about 8:00 a.m. the client called and complained that the refrigerator was not cooling. The same day on the afternoon, at the return of its work, the client called Anthony; he went to the place. He found the refrigerator compressor's motor very hot, and cycling by the electrical thermal overload.

They took the refrigerator back to the shop, verified it, and it passed all the tests. They plugged it in and it worked normally for several days. They talked to the customer; they agreed to get the refrigerator back to the place. Once there, the device was plugged in, and it started and it worked normal, but on the following morning, the client called again with the same complaint, "The refrigerator didn't cool," and it was making the same noises as the first time. They returned the refrigerator to the shop one more time, and it passed all the tests. However, before taking it to the place again, Anthony called me to consult the problem.

Questions

1. What do you think may be the problem, if any, with this refrigerator?

2. What procedure should you perform, to verify this problem?

3. The equipment had been delivered twice and showed the same trouble. Would you deliver this equipment for the third time?

4. What should you recommend doing?

5. In this case or a similar one, can you think of any other alternative that can be tried?

==

EXPERIENCE 64-A

Through my trips by the city, gas stations, car dealers, etc., I realized an existing demand in the work verification-repair of the air conditioning equipment in automobiles. Since I had taught this course, I prepared myself to obtain some benefit from it. I order the making of a placard sign, offering revision, repair, etc. of automobile air-conditioning equipment as well as window A/C unit types.

On a Saturday morning (more or less 9:00 a.m.) at the beginning of summer, I placed up the sign in front of the school.

In less than half an hour, I had a line of not less than a dozen vehicles. Moreover, at the same time, there were people with small A/C systems to be reviewed or repaired. With three students, we were working until about 2:30 p.m.

At this time the most popular refrigerant used in auto air-conditioning systems was R- 134a. We did not forget that some systems still used R-12. So, we prepared to do the job.

We also had to keep in mind the operating pressures for R-22, for the window unit's type.

Besides the "lack" of refrigerant problems . . .

Questions

1. What other problems may the window air conditioning have?

2. What the operating pressures should be for R-22, R-12 and R-134?

3. What kind of tests do we perform?

4. What do you think is the capacity, tons wise, in a regular size (4 doors sedan) auto?

5. How much refrigerant should be the charge of R-12 and R-134 for a system on the question 4?

6. What do you think some other fails (problem) in auto air conditioning could be?

===

EXPERIENCE 65-A

I was in this place (a delicatessen) before; now, it was July and the customer was having problems with three of the refrigeration systems. At this time of the year, the temperature in the basement, condensing unit's location, it was too high; usually above 90°F (32°C). Moreover, none of the three condensing units were working properly. They should never, under these so high, condensing temperature conditions.

The same day, after finishing at the school (as in Experience 64), I had already another commitment.

The three units in the basement, ranging from 1/2 to 3/4 hp had the following characteristics:

1. All of them worked with 115 volts, single phase and 60 Hz.
2. They all used refrigerant 12.
3. Although they work for different applications located at the store, they all were of the air-cooled condensers type.

Questions

Taken into consideration the high temperature in this place . . .

1. What would you do, to change the actual operating conditions?

2. What should you recommend, to make these three systems work properly?

3. How would you carry out such repair?

4. After whatever change suggested, what should be the new operating pressures?

==

EXPERIENCE 66-A

This time, the dispatcher sent me to a store located on the city's east side. This customer complained that very often, the air-conditioning equipment took a long time to begin cooling, and when it finally did, he thought the cooling effect was not as good as he expected. And at the cooling tower location (basement), sometimes water spilled on the floor.

According to him, since the A/C equipment was installed, more than two months ago, he has called the installation company; mechanics have come several times, but according to them, there is no reason for the equipment to not operate well. In other words, they couldn't find anything wrong.

I arrived at the place 11:30 a.m.

The outside temperature was 80°F.

Per the inspection to the equipment, the following information was obtained:

This system had the following characteristics:

1. A 5-ton semi hermetic unit.
2. It operated with 440 volts, 60 Hz, and 3 phases.
3. Condensing by water.

    Cooling tower located in the basement.

4. Control circuit using 24 volts.
5. A resistor surrounding the crankcase. It works with 230 volts.
6. The sight glass on the liquid line showed some bubbles.

7. It's equipped with the following controls:

    a. Timer.
       Set to start the equipment at 7:30 a.m. and to stop at 7:00 p.m.
    b. Thermostat on one of the place's walls.
    c. A dual (high and low-pressure) control. Both automatic reset types.

Other information:

1. The equipment was working.
2. Temperature at the served area's thermostat was 75°F.
3. The temperatures at:

    a. System's suction line 65°F.
    b. Liquid line 95°F.

4. Condensing water supply and return temperature differential 14°F.
5. Pressure gauges were installed, given the following operating readings:

    a. Discharge = 230 psig.
    b. Suction = 60 psig.

Questions:

1. Do you think it is necessary to connect the manifold?

2. By using the customer as well as the equipment information, what should be our next step?

3. What do you suspect may be the reason why this equipment usually took so long to start cooling and when it does it's not to the customer's satisfaction?

4. What is the explanation for the spilled water at the basement?

5. Are there any other questions to

    a. The customer?
    b. The equipment?

===

EXPERIENCE 67-A

While working with the company school's owner, in Brooklyn, and because most of the people in the office smoked, I considered the need for an exhaust fan motor and I asked for it. They sent a mechanic "electrician," who installed the required device in the flat ceiling; he went up and down a ladder many times; he was "struggling" for three days.

Watching his troubles, I finally decided to ask him what seemed to be the problem. He said, "The fan motor do not works well, and when it does, it, it is very slow." I asked him what the needed voltage was. He told me "115 volts."

I asked if he had measured the voltage value at the source; he said yes and told me it was 115 volts.

Questions

1. What do we think the problem was?

2. Do we have any suggestions?

3. What do we recommend, to fix up this problem?

4. What tool and how we use it, to recognize and solve this problem?

===

EXPERIENCE 68-A

Another incident that I remembered, it was with one of the engineers who worked in the same company school's owner. This man it was in some way, assigned to work in the World Trade Center. I overheard him to comment, in several occasions with some other engineers of a problem, about this cooling tower in which the water pipes; due to the neighborhoods constructions were constantly obstructed; cement-sand got airborne and flew going to deposit in the cooling tower causing the problem.

They had had to make serious-expensive repairs, clearing the piping. From the other engineers I heard many suggestions (filters, etc.).

Any ideas?

Question.

1. What do you think it should be done to overcome this problem?

===

EXPERIENCE 69-A

A friend of this former student, requested from him to make an installation of a refrigeration freezing equipment; in a walk-in space of 6' x 6' x 6' (216 cubic feet) located in a store of its property. The substance to be stored/frozen, were meat products (in an amount of 5,000 pounds), requiring a storage temperature of -15°F.

The product came in, at a temperature of 40°F.

The work included the calculation of the equipment size.

He came to me and requested help in this matter. I went to the place and verified the information. Once gathered such data, rather by curiosity, I make a call to a commercial refrigeration equipment's distributor, so that they could oriented me about the required refrigeration equipment's size/capacity. The contacted person made a big deal of it, and left me without any help.

Questions

1. Do you have any idea of what to do, to comply with this application's requirements?

2. What should be the capacity of the equipment for this application?

3. Do you have any other idea or contact to help?

4. In today's world; what refrigerant should we recommend to be used?

5. Once we got the installation, what should be the operating pressures?

===

EXPERIENCE 70-A

Another student (Let's called him Robert) had a commercial refrigeration repair shop. A client asked him to perform a job, in a freezer fixture, of a supermarket located at Gun Hill Avenue in the Bronx. When Robert called me, he said that in this device, a couple of companies had worked.

According to the customer, for about 1 month, the system wasn't freezing properly. The first company, make some repairs/changes in the equipment. The device, worked for few days but with the same problem not freezing. The company got back. Try for several days more, but the same result.

Another company was called; they change the thermostatic expansion valve. But almost immediately, the compressor failed. This company never came back. They even left (abandoned) in the place, some tools.

The customer says, that since the second company worked in the equipment, he noticed, at some time, that the tubing was covered with ice.

A few days later, another company was hired; they work in the equipment for a couple of days and after the compressor failed again the customer decided to call Robert, because he had heard, good references about him.

This equipment had the following characteristics:

1. A semi hermetic 4 tons capacity compressor, with an air-cooled condenser.
2. Electrical characteristics 440 volts. 3 phase, 60 Hz.

3. It worked with R-502.
4. It was equipped with a dual pressure control.
5. It worked for a freezer-counter (with a temperature of -20°F.)
6. The freezer-counter's evaporator was located at about 30' from the compressor's location.

The condensing unit was located back of the store at an attic approx. 6 or 7 feet above the ground. The temperature of this location was about 85°F.

Robert changed the broken compressor and the expansion valve again (third time); when the system was started, within a few minutes the suction line and the compressor got frozen.

According to him, he stopped the system and decided to call me; before the breaking of the compressor happen again.

I went to the place. I enter the store and saw the freezer-counter size. When I went up to the equipment location, I saw several R-502 empty cylinders.

I put to work the system, and in few minutes I felt the suction line and the compressor's body temperatures, descended until being frozen. I didn't even have the time to let the operating pressures normalize, I stopped the equipment.

After analyzing the above information:

Questions

1. What may be the cause or causes for the so-often broken compressor?

2. Why, within a few minutes of work, did the suction line and compressor got "ice-covered"?

3. What components/parts operation should we concentrate at?

4. What testing procedures might we use to verify this equipment problem?

5. For this system, what should be the operating pressures?

==

EXPERIENCE 71-A

With this Russian student (Anatoli) that since he left the school, had worked for two different companies one in New Jersey, the other in New York City, I had two (2) relevant experiences that I would like to comment:

(1). While he worked in New Jersey, he had several problems; nevertheless, this one, it was the most relevant; he was sent to a service call. The problem, according to the client, the A/C system began to operate normally in the morning; but, about 3:00 or 4:00 p.m. it stopped and it did not return to start; they made the call to request for service, but as it was too late; the company sent the service technician on the following morning.

The same problem had happened for the last couple of days. Anatoli arrived at this place at eight o'clock in the morning, and the equipment was working (started by a timer at about 7:00 a.m.), he reviewed the "entire" operation, and after almost two hours, he did not find any problem.

He called the company and made the report, and it was sent to another job.

The next morning he was sent again; he did exactly the same thing as the day before, with the same results.

The company gave Anatoli the third chance, but they told him, "If he does not find the problem, they may let him go." Then he called me and explained the problem.

I asked him three questions:

Question 1: How were the operating pressures when he took them?

He answered, "Normal," and he gave me the pressures numbers for R-22.

Question 2: What type of controls does the equipment have?

He answered, "2: high- and low-pressure controls."

I asked him, "Are they separated or together?"

He answered, "Separated."

Question 3: Do you see any printings or plans?

He answered no.

I suggested to him to look or ask somebody, and to let me know.

About twenty minutes later he called back. He had the plans with him.

Questions

1. What we may think, the problem here was?

2. What procedure or test do you perform, to solve this problem?

3. What device's function should you check?

4. What should be the proper cut-out setting on the hp control?

5. What should we look for in the printings or plans?

===

EXPERIENCE 72-A

A few months later, Anatoli call, to let me know that he was working in "Hunt Point"; the Bronx terminal market; there, he was in charge of the equipment that —— (the supermarkets chain) had in that place. He offered to help me, in case I wanted to send the students for practice.

According to him, they had a lot of equipment of all the capacities, ranging from one to 15 tons; on all the temperatures and cooling mediums (air, water, glycol, etc.).

A couple of weeks later I called him, and we started this endeavor, by sending four (4) students. This practice continued by several weeks, with very good results.

Then something important happened. A supervisor of —— was interested in the student's work and knowledge and he asked Anatoli from which school they were. Anatoli told him, and the man was interested in speaking with me. He called and asked me, if we could be interested in giving service/maintenance, to all the equipment that —— had in Long island? I answered him positively; he gave me the address so I could go, and see the place and systems. After which, I would call him to ultimate details.

I spoke to Ricardo (my son) and we went to the place; its distance was about 25 miles from the school in Brooklyn. There were all types of refrigeration systems ranging in sizes from 3 to 30 tons. Thirty systems. Some of them, worked for refrigerating walking boxes (warehouses), of up to 250 square feet. All the equipment was, in very bad shape, and worst working conditions.

They have had a terrible maintenance and it was necessary to fix everything.

Among the works to be performed: Check for leaks, pull vacuum processes to all systems, change condensers and evaporators fan motors, installing of oil separators, reviewing of all the electrical circuits, adjustment/change of controls, etc. In short, there was a lot of work for a long while. In this inspection, we expended, about 3 hours.

At our return, I called the —— supervisor and I rendered a detailed report to which there was to be done. After two days, he called me and he gave me unlimited power to initiate the works. Spoke with Ricardo and he told me about this person "Mike" retired from —— Service Company, to fill the mechanic position. With Ricardo, we went to New Jersey; bought a light closed truck for equipment, materials and tools transportation, and we began the work. This light truck does not last long, it had gear transmission problems; in New York "broke down" and was necessary to tow it and to take it back to New Jersey. We bought another one and this time this one, was perfectly used for the job.

If my memory serves me right, in this work, were used more than 3,000 pounds of refrigerants: 12, 22 and 502. Any number of fan motors, etc.

The work with the mechanic was a little difficult; he wanted to make things his way, and that, did not pleased me at all.

Ricardo went to the place, when he had an opportunity, usually at the afternoons and I only could go Saturdays and sometimes in the mornings. Most of the condensing units were located, on the roof and for me to get there was with a crane. Therefore, it was far from easy.

All the works were very interesting. But, I should mention one of the many important ones: Installation of oil separators in (6) systems 15 and 30 tons. These machines were of the semi hermetic compressor types, equipped with suction and discharge packing gland type service valves. Air-cooled condensers, using receivers; these, were equipped at their outlets with "King" service valves.

To perform the job, the refrigerant R-502 had to be recovered. I let Mike to know, so he can start the oil-separator's installation job.

All the six systems were similar in design, but three were of 15 tons and the other three of 30 tons; to make sure, I told Mike to start with, only one of the 15 tons capacities.

In order to accomplish the refrigerant's removal (about 60 lb), he used a recovery machine for 22 hours. He soldered the oil separator (connections of 1.5" diameter) to the discharge line; after installing it, he tested for "leaks" and "did not find" any. Charged the system with refrigerant and put it up to work. Two days had passed and we had a problem "a lack of refrigerant," result of a leak in one of the soldered joints of the oil separator.

I told Mike, not to do anything else with that equipment; wait until I went down there. I believe that he did not like it! It was not the first time that he demonstrated it so, when I said something to him.

The following Saturday I went to the place with Ricardo, they raised me up to the roof. I seated in front of the unit, in which the oil-separator had been changed. I study it by 2 or 3 minutes. I called Mike, he came a little reluctantly. But . . . I asked to him! Do me a favor: Look at this system, and tell me what do you see!

He looked at me surprised, like asking. What? I said to him! Tell me the parts of the equipment that you see! Not too happy, but he was enumerating to me, the different components. He had ignored the valve coming out of the condenser. I made it notice it.

Now, I asked him, "In order to remake the oil separator's solder, we must remove the pressure between the compressor's outlet and the condenser's exit. Right?" After observing a little while, he agreed. I said to him, "Now, connect the recovery machine to the condenser's service "king" valve. Make sure to close (front seat) the two valves: the compressor's discharge service and the condenser's king valves. Now let recover the refrigerant in that part of the system."

This operation was carried out until the pressure gauge reached "0" psig. It took about 20 minutes instead of the twenty-two hours that was used before. I asked him, "What do you think now? That is some difference! Ah?" He did not say anything.

Now the soldering operation could be completed.

About three weeks after that, we started working here, and hoping to give the Russian students a vote of confidence, in the month of July, we hired two of them. We raise a problem, because a few days later, on the 26th, one of them (according to Mike's report) let himself "fall" from a ladder and had to go to compensation. Thank goodness we had insurance another way; we would have a big headache with this guy. The other, we cut him before he became as smart as his partner was.

In this place, we make invoices of up to $150,000, in an approximate time of three months. I do not remember well, how long we worked there, but were by several months.

Four problems began to arise:

1. We began to notice that someone was trying to sabotage the work: we found cut pipes, disconnected controls, and a certain attitude of the people in that place against us. (We believed the people of the company that had the job before us encouraged it.)

2. Adding to this, we had to deal with a mechanic who did not want to accept suggestions.

3. The supervisor, although very satisfied with our work, wanted us to be affiliated with a union, and

4. The Russians were becoming too wise (remembering here Irina's comments about them).

With those four facts, I decided to discontinue the operation.

==

EXPERIENCE 73-A

By this time, Ricardo was working stationary, in a building located in Park Avenue and 46th. Street. According to what I understood, he was taken care of the air conditioning equipment's operation/maintenance in the entire building; in which the main client and I believe the owner; was a Swiss bank. This work kept him busy enough; for that reason, when something extra appeared that would require more time, he had to call the service company and they sent a mechanic, to make whatever it was necessary.

One of those days, Ricardo called me to do a consultation: He told to me that for several weeks, they had problem with an air conditioning equipment in a quarter of one of the main officers; the service company had sent several mechanics.

A/C system characteristics.

1. This equipment had a hermetic unit with a capacity of 3 tons.
2. Electrical features 230 volts, 3 phases and 60 Hz.
3. This system used refrigerant 22.

Besides the above characteristics, this equipment; as we must know had the other usual electrical components:

a. Selector switch.
b. Thermostat.
c. Thermal overload.
d. Running capacitor.
e. The electrical installation, of course.

This equipment was located at the ceiling of the space it served.

The first mechanic, it diagnosed that there was necessary to change the compressor that "it did not work" (Electrical problem?). A couple of days later, another mechanic came and did the change. According to, neither work, thus; he disconnects it and took it back. Another two days passed by and another mechanic with a new unit came and change it. Same result.

After three (3) consecutive unit changes (3 refrigerant loads to the air), was when Ricardo after hearing the repeated complaints of the client, called me.

Questions

Taken into account the information above . . .

1. What do you think, the problem in this A/C equipment was?

2. What tests we may perform, to verify the electrical "failure" in this system?

3. The term: "It did not work" what that means for you?

4. Do you think a job like this, had to go unsolved for so long?

5. What's the next procedure, we should recommend?

6. How much refrigerant, is the charge for this three-ton A/C unit?

==

EXPERIENCE 74-A

I believed it was by this time that the company owner of the school's leased contract for the premises on Metropolitan Avenue, was being finished and the building's owner, was transferring his business to the second floor, which at this time the school's owner had almost no use for that part of the building.

The building's owner was a very good person, and the only thing that he asked for was the removal of the A/C condensing unit; that was still working on the premises' second floor.

The company sent a mechanic, so that he carried out the transfer and relocated the equipment to the building's roof edge.

For some reason, unknown by me, for some of the company's personnel, the school was not of much affability. This mechanic that they sent seemed to belong to that group. I remembered the day he came; he entered the school with an air of superiority that made me think! This person behaved, as he was Napoleon Bonaparte! He told me who he was, and what he came for, and he asked me with arrogance.

! Do you think, that this school can teach something to me?

I did not answer his question, instead I thought *"Too foolish words, deaf ears"*; and I limited to tell (remember) him, what he had to do:

To install the condensing unit on the building's edge of the roof and to run the two equipment lines (Suction and liquid) that connected the A/C's components, between the first floor (TXV- Evaporator) and the roof (Condensing unit). After two days and already finished the transfer, I realized how an able and smart "technician" was this individual:

142

The building's front wall, where he performed out two (2) soldering procedures; it was burned, at least four bricks had been damaged and on the street's sidewalk, I believe, like 2 or 3 melted silver solder rods, were in the floor. Nor even the work of, one of my apprentice students. But good Lord!

I got an anecdote about the "techniques" of this "mechanic."

There only passed two weeks, after he had performed the job at the school, and I knew that !my technician/mechanic! Had a tremendous problem:

He was sent to the second floor of the World Trade Center (WTC) (6 in the afternoon); to repair a problem in an A/C unit (I think of 3 tons), located in the ceiling of a computer room.

The equipment had a water leak, in one of the condenser water regulating valve's connections. He closed the water (supply/return) valves and proceeded to change the water-regulating valve (WRV).

Once he finished the job, he opened the water valves and left the place.

Questions

1. Do you think there was any problem with this guy's work at the WTC?

2. Do you have any idea what the problem was?
   Probably you do not think so.

===

EXPERIENCE 75-A

I continued teaching the RSES courses in La Guardia Community College.

A peculiar event happened to one of my students who worked for one of the city's RAC service companies. This day, he arrived to the class a little behind schedule, and it trusted to me that he was a little tired, and he referred the following experience:

He was sent, by its company, to take care of a service call; in a delicatessen located at the Manhattan heights.

According to the company's dispatcher, the reason for the call was; that one of the store beverage coolers was not working. When he arrived at the place, he found that the employee who welcomed him had placed the service call; at the time, he was in charge of the store. He indicated to him the location of the equipment at the cellar.

System's characteristics: A beer cooler.

1. A 3/4-ton sealed unit.
2. Water-cooled shell-and-coil condenser type.
3. Power source 115 volts, single phase, and 60 cycles.

The system was using city water as a condensing cooling medium.

The equipment's compressor wasn't working. He found the high-pressure control was open.

The mechanic reset it the hp control and the system started but, few seconds later, stop by the High-pressure control (*) again. He installed the manifold pressure analyzer, to verify the operating pressures readings. He got a very high discharge pressure reading.

He investigated, and he found out that the "reason" this control to work (open) was, that it was not water circulation through the condenser.

(*) This hp pressure control was of the "manual" reset type.

Questions

1. What should be your approach, to this problem?

2. Should you ask any other questions to the employee?

3. What should you do to verify the "lack" of water through the condenser?

4. Should you verify the hp control settings?

===

EXPERIENCE 76-A

I believe was by this time, a former student called to make a consult; it was somewhere at the end of autumn or at the beginning of winter.

He had installed an air-conditioning system, and given service to another one in the same place; the condensing units of both of them, were in the same location (outdoors)

He said the problem was that both systems had, for several days, worked well but today the customer had called it because neither of them was working.

I asked what type of control the systems had. He said that both systems had a dual type High and Low pressure controls, and both systems were of the air cooled condensing type.

I ask to him to take a look, and tell me by which control the machines were stopped by.

After taking the operating pressure readings and watching them, he said that both were off by the low-pressure controls.

Questions

1. Besides mine, do you have any other question? If your answer is yes, what should the question be?

2. What do you think was the problem there?

3. What do you recommend doing, in order to solve this problem?

4. What tool or tools may we use to help us in solving this case?

==

EXPERIENCE 77-A

A peculiar case happened, with a school's neighbor that we had in Second Avenue. This man was from another foreign country origin, and according to him; our school had nothing to teach him.

To begin with, one day (do not remember the reason) I went to his shop; at that moment, he was starting to test a small air conditioner system for refrigerant leaks, by using the gas, from a nitrogen cylinder.

The pressure in these cylinders (when they are full), are of but about 3,000 pounds; he was going to make the pressure-test, without the use of a pressure-regulator, to bring down the pressure to a lower working and safety pressure value.

When he went to do the procedure, I told him! Stop! He asks me, why. And I told him "Wait until I leave the place, before the equipment explodes." He laughed, but he waited until I left the place.

Another experience with this "well-versed" man was:

On 2nd. Avenue and 116th. Street was at that time, a Spanish restaurant named "El Pueblo" (The Town), and this man; installed (including the electrical installation) a 5 tons air conditioning system. The restaurant's client/owner, promised him to pay the rest of the works money; the day after the inauguration (a Saturday). That Saturday night, the place caught fire and burned out.

Monday morning, my man came to the business to receive his money; all he found was the place, still smoky. According to the firefighters: The fire was produced, by a fault in the electrical installation (Very thin wires # 14), in the A/C equipment.

Here, as in the case of the —— mechanic (Experience 74) in the WTC he took the way of "Villa Diego." In plain English, "He walked away." I saw, when he was putting all his "belongings" (call them tools and equipment), in its truck and "he went" according to, the way of Florida.

I knew that, unfortunately, he only arrived at a place of New Jersey's turnpike; near Delaware because the truck's motor was burnout.

Questions

1. What do you think about this guy behavior?

2. What reference can we use to adjust the pressure testing in any A/C or R system?

3. What other precautions must we exercise when using gases to check A/C-R systems? For system's strength, and refrigerant leaks?

==

EXPERIENCE 78-A

While I worked with my partner in the school (he was always searching); I learned (I find out), like always, many things; one of them was that all the companies produced extra materials and, more important, what happened with those items.

Let me explain:

For example, a compressor's manufacturer receives an order to build, all together 10,000 units of different capacities (2,000 of each denomination: 1/8, 1/4, 1/3, and 1/2 hp, etc.)

It builds them, in a total of 12,000 compressors. The surplus (excess), with the purpose of replacement for any fault, in the 10,000 units ordered.

The 2,000 excess compressors were stored and when a unit fails, or it is returned (for any reason: mechanical, electrical, etc.), they replace it with the ones they made in excess. The surplus compressors were stored in the warehouse, among of the initial harvest.

Usually, the manufacturer's warranty lasts two years. After a certain time from that expiration date, the companies send sales representative personnel to sell all the leftover units.

To us (my partner and I), one of these representatives, offered 5,000 units, at the cost of $10.00 each one. Of course, we did not have such amount of money, but companies like "No Name" bought them and stored them; then they sold them, at retail or in groups, to the present-day price in the market. (Example: A compressor of 1/8 hp was worth $80.00.)

No Name had branches in other states of the union as well, e.g., in Miami, where some buyers came from South America, the Caribbean, etc. They bought those products with a discount of 25% of the regular present-day price.

Many of those units were in a good operative condition, but some have defects and when installing them, they show those defects. I, personally, had the experience (here and abroad), as a buyer, as well as a consumer; with several units that I bought, or got in these warehouses.

This one was kind of personal experience; it did not happen to me, but to one of my former students: He bought a 3/4 of a ton sealed unit. When he went to use it, after he installed, the compressor was stuck. He tried everything in his knowledge, but nothing worked. So he disconnected and carried back to the store. The attendant asked him if he was sure, and if he knew what he was doing; trying with that to avoid the unit's replacement. But our student knew better, and he had good arguments to get the replacement.

Let's see how we can acquire something from this . . .

----

1. How do you know, when you buy one of these compressors, if it is new or is one of those obsolete ones or "doubtful" from "surplus"?

2. What should you do to verify the fabrication date of one of these devices?

3. What do you think was the "argument" that the student presented to get the replacement?

4. What can we do to verify the device condition before "wasting" our time and money?

==

EXPERIENCE 79-A

By my activities in the school and other shores, it was not always possible for me, to go and help my former students; when they required me to. Therefore, many times, they have to complete their works by themselves; most of the times they did it with good results, but in some occasions the works did not came out; the way they intended-desired to.

This student (very stubborn) installed a 5-ton air conditioner system in a restaurant.

Other characteristics of this system:

1. A self-contained A/C system with a water-cooled shell-and-coil-type condenser.
2. Electric characteristics: 230 volts, 3 phases, 60 Hz.
3. Using refrigerant 22.

Ambient temperature = 80°F.

At a distance of 40 feet of the A/C unit, he installed the cooling tower and made all the piping installation. When he went to run the equipment, it did not work well; then, he decided to call and consult me.

I went to the place; when I put the equipment in operation, the cooling tower was but a "steam" maker. I verify the installation; and found out that the water pipes were, not big enough to carry out the water the equipment required to operate.

He had used copper tubing of 5/8" diameter. I explained to him that a system like that required much more water, than tubing that size could carry out. My explanation wasn't enough, and in his stubbornness, he decided/preferred, to return the money and to cancel the transaction.

Questions

1. How much water needs the condenser in an A/C system of this size?

2. When in doubt about the size of tubing/piping required in an application, what do you think should be a good reference that can be used?

3. What should be the normal temperature difference when using a cooling tower?

==

EXPERIENCE 80-A

Another of a former student (Robert) works. By several days, he was trying to get back to work a food-cooler device, located in a supermarket. According to him, before they called him, there were already, two mechanics from two different service companies, but they could not find the problem either.

They all recommended changing the unit.

In this equipment: The Potential relay, the starting capacitor and the thermal overload had been changed three times. The place was a cemetery of these devices.

Originally, the customer said, that the first mechanic found the starting capacitor blown out.

Equipment characteristics: Split-type equipment, where the fixture box was located on the first floor, and the condensing unit at the store's cellar.

1. Semi hermetic unit using 230 volts, single phase and 60 Hz.
2. Water-cooled condenser with city water.
3. Used refrigerant 12.
4. Compressor's motor of 3/4 hp.
5. Uses an overload, a potential (voltage) relay and a starting capacitor.

According to Robert, all of this semi hermetic unit windings resistance readings values, were normal:

1. Starting winding (15 Ω).
2. Running winding (1.5 Ω).

3. Sum of the two windings (16.5 Ω) and no "ground" reading.

He had already changed the potential relay, the thermal overload and the starting capacitor. Nevertheless, the unit started but after a few seconds, it stopped by action of the thermal overload.

I went to the place and indeed all the readings were normal. I energized the equipment, it started but the LRA reading remained for too long so the thermal overload kick-out disconnecting the unit.

All the relays and capacitors found in the place were exactly the replacement for the specific unit; so they were compatible. Besides, all the relays were tested and were OK.

Questions

1. What do you think was wrong here?

2. What was making the thermal overload, to stop the machine?

3. Why the LRA remains so high, for so long?

4. Besides the tests been already performed, what other tests should you recommend?

5. Should you recommend changing the compressor, as well as the others, did?

6. There was some other procedure that can be tried?

===

EXPERIENCE 81-A

It was late in the autumn, a former student (Mark) consulted me about another problem, this time; a commercial refrigeration fixture, in a store (a display cabinet for cooling meats) in which, Mark said, a couple of mechanics had worked, a week or so before.

Now the client had called him. The system's fixture was located in the store; with the condensing unit located in the cellar. According to the owner, after the first repair, change of the low-pressure control, pulled vacuum and charge R-12 the machine was working well.

A couple days after the repair, the client said, the stored meats (Desired temperature 15°F.) not always received sufficient cooling. We went to the place, through a stairs-way we got to the cellar (these space was very cold, there was no heating) the temperature in the place was 30°F the equipment was not in operation, but the compressor was "lukewarm."

The system had the following characteristics:

1. It was a 3/4 hp unit with an air-cooled condenser.
2. It used R-12.
3. It was provided with a low-pressure control.

The pressure gauges were connected, the pressure readings were taken; the system's low pressure (R-12) reading was 28 psig.

From the information above and our experience.

Questions

1. Any suggestions about why the machine was displaying this problem?

2. What procedure do you recommend to perform to remove the problem?

3. Which should be the most important tool to be used here?

===

EXPERIENCE 82-A

At the beginning of November of this year (2003) there came to our school a teacher of a city's public school in Queens; for the repair of a refrigeration machine, used in a decorative fish tank. Somebody spoke to him about us.

According to him, two refrigeration companies had refused to repair it (he said because they alleged something about not knowing "the water temperature requirements").

The piece of equipment that he brought had the following characteristics:

1. A ¾-ton sealed unit, with 115 volts, single phase, 60 Hz electric features.
2. Attached to the compressor's body was a suction-service valve of the packing-gland type.
3. Unit using a potential (voltage) relay, a starting as well as a running capacitors and a thermal overload.
4. A water-cooled condenser of the shell-and-coil type, equipped with a packing-gland service (king) valve at its outlet.
5. The equipment used R-12.

We verified the sealed unit, which we found was "burned out."

The technical characteristics were taken, and that information was given to the teacher; so, he could get the right replacement. We also recommended, changing the electrical items cited at #3 above.

After disconnecting the faulty compressor, the equipment's condenser unit was submitted to the following procedures:

a) By using first refrigerant 11, nitrogen we proceeded to decontaminate it.
b) When the new compressor came, after we verified for proper replacement, it was installed.
c) The entire equipment was checked for leaks; water and refrigerant.
d) Pulled a deep vacuum of about 3 hours.
e) Following the plate's information and our experience in the matter (at that time, we had a decorative fish tank in the school, as well as one at home), the proper system's refrigerant charge was calculated (by weight) and added to the equipment.

Finishing our job, as we did not have, the other part of the system, we could only instruct the man so that, he try to decontaminate the other part of the system (liquid line, liquid control, evaporator and suction line) in his place; changed the liquid line filter-dryer device; connect this condensing unit (appropriately), check for refrigerant leaks, pull a good vacuum to the rest of the system, add refrigerant if needed, and after those procedures; put it to work, checking both amperage (LRA & FLA) values.

Verify the suction, as well as the liquid line temperatures. Then, after running the system call, and let us know.

The day after, he called us and said that the equipment was working very well. He told me, that he did not even have to add refrigerant.

Question

1. How can we get the information, about the required water temperature for a system like this?

2. How much refrigerant, by weight, had to be charged to this system?

===

EXPERIENCE 83-A

This service call was to a bar located at 2nd Avenue and 36th Street. The customer's complaint was that the air conditioning was not working properly for some time, since several mechanics from a couple of service companies had been working and making some changes on it. The entire system was installed indoors.

A/C characteristics:

1. It was a split system with a hermetic 7 1/2-ton compressor.
2. Remote air-cooled condenser.
3. Used refrigerant 12.
4. It was equipped with a receiver. At its outlet, a service "king" valve.
5. It worked with 440 volts, 3 phases, and 60 Hz.

I got familiar with the system and noticed, that it had been equipped (seems recently installed) on the suction line; with a suction pressure regulator (SPR).

Usually, this regulator is used; to limit the pressure in the crankcase (for me, no use in a system like this).

The compressor was stall it, but lukewarm, which indicated, that it had been working. I proceed to install the manifold pressure gauges; the low-pressure side hose, on the available packing gland service valve attached to the unit. The high-pressure side hose, to the king valve at the receiver's outlet.

By using the voltmeter, found the low-pressure control (installed on the service suction valve, adjusted cut-out 28 and cut-in 42 psig), it was open.

The actual suction pressure reading was low (32 psig), notice as well (sensing by hand), a small temperature difference across the SPR device. A few minutes passed, the system started, but, within a couple of seconds stopped again.

Questions

1. Why do you think, the SPR device was installed in this A/C system; charged with R-12?

2. What do you think or suspect, what the problem may be?

3. What should we do to fix the problem?

==

EXPERIENCE 84-A

This service call was to a warehouse (Bodega), located at Broadway Avenue and 208th Street. The customer's complaint: System was not working! The condensing set (Copeland compressor) was located at the basement; it worked, for a commercial refrigeration system display box located on the store.

Equipment characteristics:

1. The ¾-hp compressor was of a semi hermetic type.
2. A thermal overload.
3. A potential (voltage) relay.
4. With starting and running capacitors.
5. ID plate information: 120 volts, single phase, 60 Hz.
6. Refrigerant 12.
7. Air-cooled condenser.

On the wall, there was a disconnecting box, equipped with a handle (OFF positioned) and one burn-out (plug type 30 amps fuse) the other line (supposedly the "Neutral") went directly to the compressor's junction box.

Before conducting the electric motor's test, and to make myself safe and sure, by using a voltmeter, the voltage supply readings at the compressor's terminals, were taken and the meter registered a 115 volts reading!

What! The nameplate said 115 volts, the lever at the fuse box were disconnected, and the fuse was burned out.

This was simple! Unbelievable!

I disconnected the live line from the motor; as well as the relay, the thermal overload and the starting and run capacitors. Then, by using an analog ohmmeter scale "X1" Ohm (Ω -1) adjust the meter to zero reading then, the following procedure were performed:

a. When connecting the electric instrument testing prods to the motor's terminals; the meter gave a "continuity" reading in each of the three tests: C-S, C-R, and R-S. This indicated a "short circuit" between them; it was *burned out*.

b. Just to make sure; the "ground" test was performed as well, giving "positive."

To order the appropriate replacement; I proceed to take the information from the Copeland compressor's Nameplate: Model and Serial numbers and the rest of pertinent information.

Questions

1. In this electrical installation, normal conditions, what voltage reading value should be obtained; at the unit's junction box?

2. What do you think, about the information obtained with the voltmeter?

3. What kind of investigation should you conduct?

4. Should we ask other questions to the customer?

5. Which those questions should be?

==

EXPERIENCE 85-A

For this service call, (I was sent by mistake). The problem was for a Domestic Refrigeration application; a household "Refrigerator." Nevertheless, in the past I saw the same problem that may appear today as well, in any Domestic, or Commercial Refrigeration fields in the "Medium" and "Low" temperature applications; since the operating pressures in these applications, can have similar operating characteristics in both fields, with anyone of the refrigerants used today.

Note. This problem will never appear on high temperature application.

Equipment characteristics:

1. It had two doors (freezer on the left side). Used refrigerant 12.
2. 115 volts, single phase, and 60 Hz.
3. It had a capacity of 1/3 hp.
4. The condenser with forced air-flow was located under the device.
5. The system was equipped with a defrost timer, which was located at the lower-back part of the refrigerator.

Particularly, the complaint, according to the domestic servant who was in charge: "The refrigerator it worked well; then, suddenly defrosts, remains likewise until all the ice on the evaporator's surface disappears; then it returns to freeze again." The same operating condition, repeats in the same way, over and over again.

Moreover, according to this lady, who had worked in this place for the last four years, until a couple of days ago, the system has never presented or shown this problem.

At the time of the service, although the compressor was working, the evaporator didn't show frost; neither did it make a sound indicating the passage of refrigerant through the capillary tube, nor boiling in the evaporator.

Note: Remember the information on the first paragraph.

This problem can be showed in any refrigeration applications of the low and medium temperatures.

Questions

1. What do you think, it was happening here?

2. What procedure should we perform to find the problem?

3. What should be the recommended repair?

4. What should be the logical explanation for this problem?

===

EXPERIENCE 86-A

By information from a student, I found out about an A/C job in a gifts warehouse, near Times Square. In agreement with the store's owner, the building provided chilled water during the summer season, through a copper coil located at an upper location of the store.

This coil had been repaired many times, because in the winter they forgot to drain it, as a result the water got frozen and broke the coil until at present time, it's not possible to perform any more repairs. Now, the only solution was to change it.

I took the coil measurements and tried to find a replacement. I spent several days attempting to locate one, but it was impossible. It was necessary to make it, by order.

With this information and the size of the copper coil, I got in touch with some companies and the cost but the cheapest one was of $2,500. I spoke with the owner and we arrived at a repair price of $3,200. He accepted. I was ready to call and to order the coil manufactory.

Questions

1. What should you do in this case? Not much. Right?

===

166

EXPERIENCE 87-A

This student found discarded in the street a brand-new 9,000-Btu air-conditioning equipment; he took it to his house/shop; submitted the unit, to all types of tests; to verify its electrical condition. He said, with "perfect" results. Never the less, later he connected it to the 115 volts source of electricity, the fan-motor ran well in both of its speeds but, the compressor's motor, never started. A few seconds after being energized, it kicked-out by the thermal overload. He called to the school and requested permission to bring it in, such permit was granted.

In the school's shop, the equipment was placed on one of the work-benches; before showing the students a verifying procedure, I started by explaining them, the device's electrical characteristics.

System characteristics:

1. The compressor's motor (3/4 hp, PSC type)
2. The fan's motor (1/6 hp, PSC type) (CCW), two speeds.
3. It worked with 115 volts, one phase, 60 Hz.

Air-conditioning systems, from 3,000 Btu (1/4 of a ton) until approximately 60.000 Btu (5 tons), use the permanent split capacitor (PSC) type of electrical motor connection, and use the following basic electrical components:

1. Motor-compressor
2. Fan motor (usually several speeds)
3. Selector switch
4. Thermal overload
5. Thermostat
6. Running capacitor

(Usually, but not always, they had an electrical diagram.)

The bigger the system, the more electrical components are added.

Questions

1. What do you think may be the problem?

2. What kind of tests should we perform?

3. On equipment like this (under normal conditions), what should be the resistance readings values, when taking them from the equipment's connecting plug?

==

EXPERIENCE 88-A

In many occasions I went to United, a store that carried all kinds of stock for R & A/C repair parts; they were located at 125fh Street between Second and Third Avenues in Manhattan. By talking with the attendant, I found out that in the second floor of the same building was a school (OIC). I learned, as well, that they were having problems with the air-conditioning equipment; they were looking for somebody to fix it.

I went up to the school's office, and spoke with the director, who very emphatically said to me that the compressor of the 15-ton equipment "was burned out," according to a company to which he had consulted. I offered myself to find out the price of the replacement, as well as the cost of the repair.

However, before doing what I offered to him, and to be completely sure; I asked permission to verify the entire system's condition. Since I was in his office, and I saw the thermostat; by curiosity, I check it; it was settled to the lowest adjustment.

To give the asked permission, he was a little reluctant (I believed, influenced by what the mechanic-company had previously informed him) but, anyhow, he allowed me to do it.

I let him know, that should be no compromise from either part.

Then, I went to the location of the system, on the roof.

The motor-compressor (A/C equipment) had the following characteristics:

1. It was a semi hermetic W-type 15-ton unit.
2. Equipped with Discharge & Suction packing gland service valves.

3. Air-cooled condensing type.
4. It worked with 440 volts, 3 phases, and 60 hertz.
5. Used refrigerant 22.
6. It used a dual pressure control.

Questions

According to what we see,

1. What kind of tests should you carry out?

2. What type of tools we should use to perform such tests?

3. What do you expect to find out from this investigation?

4. Which two types of repair arrangements, if any, can you offer?

5. In this case; can we recommend the repair to be performed here?

6. If the answer is yes. Which should be such repair?

===

EXPERIENCE 89-A

Whenever the R & A/C technician is using any gas (nitrogen, oxygen, carbon dioxide [CO_2] etc.) or any one of the refrigerants in the market). It is highly recommended to verify the true identity of the gas being used by *all* means available (pressure, color codes, chemical formulas, connecting cylinder's threads, connecting adapters, etc.).

This experience took place in the first floor of a supermarket. The mechanic changed a hermetic compressor and to verify for refrigerant leaks; he used a compressed gas in an unidentified cylinder; so, the content of this gas was not clear (it was no color code and the connections, were not those of a usual nitrogen cylinder) the mechanic, erroneously "adapted" the connections, and without using the appropriate regulator; he pressurized the system.

As soon as the gas got into the system, the compressor exploded and its upper part went through the ceiling; going to end in the market's second floor. Luckily, there were no personal misfortunes.

Questions

1. What do you think of this "Accident"?

2. What should we do to prevent accidents like this?

3. Instead of using nitrogen to perform a leak test, which other gas can we use; to do a refrigerant's leak test?

4. What should be the test pressure value used in the system's high-pressure side?

5. What should be the test pressure value used in the system's low-pressure side?

===

EXPERIENCE 90-A

This service call was for a butchery store. Customer complaints: "Refrigeration system has to be refilled with refrigerant too often," he said that several companies had checked the system but; no one found any refrigerant leak.

One of these companies, even recommended; buying a new entire unit, because this one (according to them); was too old.

According to the customer, the system was about eight years old, and this problem started five or six months ago.

The refrigeration system had the following characteristics:

1. It was a ten 10-ton capacity system. Low temperature (-15°F.)
2. The compressor was of the semi hermetic type.
3. The power source was of 230 volts. Three phase, 60 hertz.
4. Used refrigerant 12.
5. It had a water-cooled condenser of the shell-and-tube type.
6. Condenser water supply from a cooling tower.
7. On one end, at the top of the condenser, a pressure relief valve was connected.

Note. *It is* always *a very good idea, to talk to the customer and hear from it, about the complaint's "history." This way, the information may help us to save some time by not wasting it in what others had expended no only that, but materials as well.*

By doing this, we may look in those places, items that the others did not.

That's one of the secrets of fast troubleshooting!!

Questions

1. After so many people, looking for the same problem; do you have any idea or comment about where this leak may be located?

2. If you were attending this service call and wanted to avoid wasting time by using the problem's "history," what procedure should you perform?

   In what place or places should those procedures be carried out?

3. What *simple* procedure or procedures can we conduct to find that leak?

EXPERIENCE 91-A

This was at my beginnings here. At this time, we were working in a building. The last tenant was moving to another place, and the building's owner was preparing it for new tenants.

The company assigned me to work with a mechanic; who was a very nice person. Our work was to remove all the air conditioning equipment (100 tons per floor) with capacities ranging from 3 to 7.5 tons; this building had more than 20 floors (but like many other buildings in New York City, it hadn't a floor 13.)

I did not know exactly why, but the working form here seemed "rare" to me, and I think to anyone as new as I was (I would call it "abundance" or "wasteful"). Let explain myself: In order to vacate this place, and to prepare it for the new renters, everything that was in the building went for the sweepings.

Armed with acetylene cutters six feet long, they cut the chains from which the lighting fixtures hung, threw away bookcases, cut water pipes, and removed all the electrical installation; in essence, everything. What remained of the building were the single bare walls and the columns that held it.

For the new air-conditioning equipment, they installed two 100 tons condensing units; every other floor (cellar, second, fourth, sixth, and so forth). For this effect, were several companies in charge of the work. It called my attention, the form they worked.

Example: In order to verify the A/C system for refrigerant leaks, they used a vacuum procedure. Due to the size of the equipment/ system, they put a vacuum pump to work day and night for several days. Then, they stopped the vacuum pump and waited for a while. If they noticed the vacuum had diminished, they assumed, it was the indication of a leak. Now, after all this waste of time, they started to perform the right procedure to locate the leak or leaks. This was an "upside down" or "reversed" form of work. I don't have the proper describing noun.

They had enough problems, especially with a leaking system in one of the floors. After several weeks, I noticed that they worked and worked and they did not arrive anywhere. I commented to the mechanic with whom I was working. His response was a shoulder's shrug, and he said that that was their problem, not ours.

I notice, that in several occasions, they charge the system with refrigerant without finding the refrigerant leak.

As I mentioned, all the systems used R-22, meaning a refrigerant charge of at least 300 pounds.

Note: Here, as in *all* my experiences, I only say what I saw. I'm not trying to affect or judge nobody. After all, this happened a while ago.

Questions

1. Can anybody please explain to me this working form? Because for me, the regular refrigeration system was and still is the same; and the procedures used then are exactly the same ones that we use today. Probably, today we are more aware of the right way of; performing the procedures.

2. What sequence should you recommend to perform the above job?

===

EXPERIENCE 92-A

With a different mechanic, we went to another service call, this time to verify an A/C system condition. Another service company served the system, but now, the customer wanted to make a service-maintenance contract with the company for whom we work.

The A/C system had the following characteristics:

1. A 25-ton system, with a W-type semi hermetic compressor.
2. Three-phase motor, 440 volts and 60 hertz.
3. Working with R-22
4. With a water-cooled condenser, water supplied from a cooling tower.
5. Equipped with a dual-pressure control.

Note: Here, there were two similar systems in the same location. One was in operation. The other, the one we were going to service, didn't. They both had the same characteristics.

According to the customer, the mechanic of the last service company was trying to perform some procedure that required isolating the compressor. We found that the front-seat packing-gland-type service discharge valve was closed.

We also found that the high-pressure control still jumped out, we assumed, to allow the system to run without this safety device. Because of this, the discharge pressure went so high that one of the compressor cylinder's heads flew off, getting inserted in the ceiling overhead. Luckily, no damage was produced other than the compressor itself.

Questions

1. What do you think about this testing procedure?

2. Could we name some of the processes in which we must close, front seat this packing-gland discharge service valve?

3. What precaution should we exercise when the discharge service valve, has to be front-seated?

==

EXPERIENCE 93-A

This was a regular routine maintenance service to one of the various (different temperatures) commercial refrigeration systems (with water-cooled condensers) working in a delicatessen.

The customer talked to me, and among others things, he asked me if I had any idea why for the last several months, he had noticed the water bill was kind of too high. I told him that I was going to verify all the systems operation, and I may find an answer for his worries.

Systems characteristics:

1. Four units ranging between 1/2 and 3/4 hp.
2. All used 115 volts, single phase, and 60 Hz.
3. They all used water-cooled condensers equipped with WRVs. City water source.
4. They all used R-12.
5. They all were equipped with dual-pressure controls.

Questions

1. What kind of procedure should be performed to verify if the cause of this problem is something in the refrigeration system's operation?

2. What type of tools should be used to investigate this problem?

3. When using city water, what difference in temperature should be expected?

4. With this type of cooling medium, in normal conditions, how much water can we expect to go through the condenser?

5. What procedure should be performed to identify this type of problem?

6. What should be the normal discharge operating pressures?

===

EXPERIENCE 94-A

This service call was to the Polyclinic Hospital, located in 52nd Street between 8th and 9th Avenues in Manhattan. At the administration office, the person in charge said that since the previous day, a mechanic serviced the air-conditioning equipment. It worked for a while, then completely stopped working.

The characteristics of the equipment were:

a) It was a sealed unit type, 5-ton capacity.
b) It used 220 volts, three phases, and 60 hertz.
c) A circuit control (24 volts).
d) Used R-22.
e) System was equipped with a dual-pressure control type: high-pressure control (manual reset), low-pressure control (automatic reset).
f) It was a split system, with the two component parts: one located in the office (TXV-evaporator), the other (condensing unit) outside the building, with a distance of about 15 feet between each other.
g) The two fan motors (evaporator and condenser) of the PSC type worked with 230 volts, single phase and 60 Hz.

Most of the electrical information registered above was found in the equipment's ID plate and electrical diagrams.

The handle in the fuses' box was in the upward position (closed "ON"), in the box, two of the three 30-amp fuses of the cartridge type were burned out.

According to the investigation, in the control circuit it was found, the hp control (manual-reset type) "tripped out." A wood chip was also found to be holding close the contactor's mechanism.

I guessed the mechanic was not able to find why the compressor was not running, and he decided to force it to work by inserting the wood chip in the contactor.

Questions

1. What do you think of this practice?

2. What should be the reason for the original "NO A/C" system's operation?

3. Other than the tests already performed, what others should we carry out?

4. What should be the next procedure to be performed?

===

EXPERIENCE 95-A

A couple of weeks had passed, and again there was another service call to the Polyclinic Hospital. This time, according to the front desk, one of the boxes where they keep corpses was not working well. They asked me to wait until the person in charge of the equipment showed up.

When this person arrived, with a peculiar smile on his face and without a word and with a signal of his right forefinger on his right shoulder, he indicated to me to follow him.

So I did. We took an elevator, and we went to one of several basements that the hospital had. When we arrived at this basement, we continued walking for about 100 feet until we got at the front of a wall of drawers (six of them) of about 24" x 24" each.

This man stopped in front of one of the drawers, pulled one of them out, and one of the corpses in them was exposed. Immediately afterward, he slapped with the palm of his right hand the cheek of the deceased, and at the same time he says to me, "Not cold enough. See?" And with a gesture, he persuaded me to do the same. I said to him, "No, thanks, I believe you."

I walked a few more feet, and at the end of the drawer's wall, at the end of a small corridor (next to the boxes) was where the condensing unit was operating.

Equipment characteristics:

1. It was a semi hermetic type air-cooled condensing unit with a 5-ton capacity.
2. Electrical features: 440 volts, 3 phases, and 60 hertz.

3. A receiver, located under the compressor mounting base, with a king service valve at its outlet.
4. System working with R-12.
5. It had a dual (high and low) pressure controls.
6. On the liquid line, a liquid sight-glass indicator was used.

The temperature, at the location of this equipment; was 78°F.

Both of the lines' temperatures were suction +/-65°F (warmer) and liquid +/-85°F (colder) than normal.

I returned to the truck to get the toolbox and the implements needed to solve this problem.

Questions

1. What procedure do you perform to find out the cause of this problem?

2. What other questions should you ask this man?

3. What should be the operating pressures for this system?

==

EXPERIENCE 96-A

This experience took place while I was teaching at another school. To carry out the practice part in the workshop, they brought several used air-conditioning units, with capacities ranging between 3/4 to 3 tons (9,000 and 36,000 Btu).

It caught my attention, and let the students to know, of one (1 1/2 tons) system with a working voltage value of 230 volts, which seemed brand new (the cover was absent). Without any preparation; I just thought that it was a very good opportunity; to show the students the use of the analog ohmmeter, by checking the electrical unit's conditions.

On a blackboard, I drew the A/C system electrical diagram and components.

Then, with the participation of all students in the class, I proceeded to select in the ohmmeter the scale of x 1 ohms (x 1 Ω).

Note: *We make sure the selector switch was set on the OFF position (open); and the thermostat on the ON position (closed)*

After adjusting the ohmmeter instrument reading to zero "0," the following procedure was carried out:

1. The instrument's leads were connected, to the two prong pins on the A/C equipment inlet connection.

2. The ohmmeter gave a reading of "∞" infinite (No reading resistance)

3. Step by step, the selector switch was moved; to each of the following positions:

    a. Fan "Low speed," the meter register a reading of 35.0 Ω

    b. Fan "Medium speed," the meter register a reading of 28.0 Ω

    c. Fan "High speed," the meter register a reading of 24.0 Ω

    d. "Cool" and fan "Low speed" instrument register 11.9 Ω

    e. "Cool" and fan "Medium speed" instrument register 10.9 Ω

    f. "Cool" and fan 'High speed' instrument register 10.3 Ω

4. The equipment was verified for "ground" as well, by using the ohmmeter scale of 10,000 ohms (x 10 KΩ),

Just as a demonstration and to show the students the problem, the equipment was energized.

Results: *The fan motor worked, on its three speeds well; Amperes readings were OK. But when the thermostat was closed, and the compressor was energized, it did not work, the starting amperage remained high and within a short time, it tripped-out by the thermal overload.*

Questions were asked to all participants about the reason or reasons for the faulty A/C equipment operation.

No answers.

The equipment was de-energized.

The electric motor of the sealed unit was disconnected. Its windings were tested, and the ohmmeter gave the following resistance readings:

5. Between the unit's terminals:

a. Common "C" and "R" Run _____ ___ 2 Ω___

b. Common "C" and "S" start _____ ___18 Ω___

c. Run "R" and "S" start _____ ___21 Ω___

Questions

1. What do you think the problem in this A/C system was?

2. Why do you think is this A/C equipment here?

3. By reading all the Ω numbers ("a" through "f") step 3 (first page) and looking at the unit's reading windings ("a" through "c") step 5, *do you agree with the resistance readings ("d" through "f") on step 3?*

4. What do you think should be the readings ("d," "e," and "f") on step 3? (first page of this experience).

===

EXPERIENCE 97-A

For this experience, this man came to the school and asked if we could verify a freezer that he had in his house. He had purchased it only two months ago at a very convenient price.

I asked him to bring it, so we could verify it.

Two days later, he brought the freezer. This was in a very good shape; it looked new.

With the students, we proceeded to check it:

It had been manufactured in a European country.

The ID plate had the following characteristics:

    1. 115 volts   2. Single phase   3. 50 cycles.

We tested the unit for resistance (ohms), and it showed a short circuit. It was burned out.

The ground condition was verified as well; it also gave a positive reading.

According to the customer, at first it worked excellently until it presented the condition. It stopped cooling, and when it was energized, the fuse burned out, and electricity passed through it.

Questions

1. What was the biggest problem you found from the above information?

2. What do you think was the problem with this equipment?

3. What explanation should be given to the customer about the problem and the repair?

4. What should you recommend being the appropriate procedure to carry out in this case?

==

EXPERIENCE 98-A

It was at the end of spring, and for some reason I had to call a former student of mine to ask him about something. After the ring, he answered me, and after greeting I asked him where he was.

He said that he was near Pittsburg, Pennsylvania, in a factory (preparing the A/C equipment for the coming summer) that he trusted to me, that in this system (and time, the last year), he had the same problem. It went on for a couple of years, this lack of refrigerant; they couldn't find any leaks. And now, he had to add refrigerant because it was not sufficient again. Otherwise, during the rest of the time, the system ran without problems.

I asked him about the characteristics of the system: He said:

1. It was 30-ton system with an open reciprocating compressor type, equipped with packing-gland type suction and discharge service valves.
2. Air-cooled condenser, at its top a pressure relief valve.
3. A receiver (with a packing gland "king" valve at its outlet) located below the condenser.
4. It used three thermostatic expansion valves for the same number of evaporators.
5. To control the space's temperature, each evaporator had a solenoid valve in each of their liquid lines.
6. Equipped with high- and low-pressure controls.
7. Used refrigerant 22.

Then I asked him if now he had found the leak, and he told me no. Then I asked him a couple of questions more:

1. Did the machine work all year round? He answered yes.

2. What did he or his mechanic do the past autumn or winter to keep the equipment working? He did not understand that question.

3. In the past (fall or winter), did he or his worker add refrigerant to maintain the discharge pressure at a normal working level? He answered yes.

4. At the beginning of the last (summer), did he remove the excess of refrigerant? He answered no.

Questions

1. According to his answers, where do you think the refrigerant's leak, if any, can be located?

2. Do you have any suggestions to help in solving this problem?

3. What procedure should he carry out to find the leak?

4. What precautions should be taken to prevent this problem again?

===

EXPERIENCE 99-A

For this service call, a supervisor told me that for some time this air conditioning system was burning out the solenoid valve's coil, located on the liquid line. They had changed the valve's coil in several occasions. He asked me to go, investigate, and take the appropriate action. He also suggested, when I got there, to get in touch with the place's engineer.

The place was a hospital located on Sixty Street and Ninth Avenue. I arrived there and went to the engineer's office. He was waiting for me, and he gave me the same information as the supervisor.

I asked him how long this was happening. He said, "About six months since the system failed to work and a mechanic from another company changed the coil."

We went to the location of the problem valve.

When checking (by touch) the valve's cover, it was found to be too hot; and it was vibrating more than these types of valves usually do. We asked, how often the coil burned out and if he had one of the burned ones; he answered almost every week and showed me not one but two of the coils. He said that they try not to work the equipment, afraid of the problem.

I read the information on the coil's bodies; they were identical. I verified the electrical connections; they were tight and right. I also measured the voltage supply, and it was right (230 volts) according with the valve's ID information.

With the knowledge that we had of these devices, answer the questions below

----

Questions

1. What do you suspect produces the solenoid's coil burning?

2. Why were the valve's body so hot and the vibration so intense?

3. What do you think was the problem?

4. With the above information, what else should we do?

5. What tests should we perform?

6. What should we recommend?

==

EXPERIENCE 100-A

By the year 1992, when the Environmental Protection Agency, proposed the examination for the "Proper Use of Refrigerants," all organizations offered its affiliates a preparation course to discuss and inform all the necessary requirements to obtain this certification.

The Refrigeration Service Engineers Society, with which I'm affiliated since 1968, offered seminars to provide that kind of instruction. In order to provide this course for the Spanish-speaking members, they asked me to teach such seminar.

For this purpose, they gave me a copy of the RSES test (translated to Spanish) and information with the EPA requirements. The problem was that the translation from English to Spanish was made by electronic means. Like any translation by this means, it was very literal, and it was quite incomprehensible. So I let them know, and I made my own translation.

I made the translation and the corresponding examination. The attendance to the seminar was pretty good. About fifty people attended. The test was provided at the end of the class.

The results were sent to the RSES headquarters, and a week later, all the students obtained their certification.

Talking about these seminars, in the year 1994, RSES, through La Guardia Community College Continuing Education Program, offered a seminar at this location. They asked me to get in charge of this course.

The EPA certification in English had an attendance of twenty-five RSES members.

As usual, the instruction was given, and in the end the examination took place. Results were sent to RSES headquarters, and a couple weeks later, the completion certificates were issued to the participants.

===

EXPERIENCE 101-A

This service call was to a supermarket. The refrigeration equipment had two walk-in rooms. One was for a low temperature (15°F.), used for meats products. The other was for medium temperature (48°F), used for milk products, vegetables, etc.

According to the client, the medium-temperature walk-in room was not cooling.

Temperature at equipment's location was approximately 75°F.

The refrigeration equipment had the following characteristics:

1. Semi hermetic unit 2 hp, 230 volts, single phase, and 60 Hz.
2. Two thermostatic expansion valves.
3. It used refrigerant 12.
4. Condensation by water (cooling tower).
5. Both rooms were equipped with solenoid valves in their liquid lines.
6. The medium-temperature evaporator had in its outlet a pressure regulator (EPR).
7. Both sight-glasses on liquid lines were "full."
8. A dual (high and low) pressure control.
9. The liquid line on low-temperature application was a bit warmer (approximately 85°F) than normal temperature.
10. The liquid line on medium-temperature application was at ambient temperature.

The manifold was installed, and the operating pressures were verified; both were normal. Suction = 9 psig. Discharge was approximately 125 psig.

Low-pressure gauge was installed at the (EPR) outlet of the medium-temperature walking box. It was founded steady at approximately 65 psig. The regulator was a little warmer.

It also was verified:

1. The two liquid lines' solenoid valves were energized and in normal operation.
2. The superheat in both thermostatic expansion valves applications were:

> Low temperature was approximately 9.5°F. Normal.
> Medium temperature was about 68°F. Too high.

Questions

1. Other than the ones performed, what other procedure should you do?

2. What should your diagnosis be?

3. What should be the pressure at the medium temperature evaporator?

4. In normal operation, what should be the normal temperature at the suction line?

==

EXPERIENCE 102-A

A former student, "Paul," came to the school asking for some help. For some time, he has been working as a maintenance technician of refrigeration and air-conditioning systems for a chain of fast-food restaurants, McDonald's style.

According to Paul's information, the previous person in charge, probably looking for economy as well as efficiency, changed in one of the restaurants the cooling tower (open type) cooling system from water to glycol.

Ever since Paul became in charge, there have been several complaints on the place's cooling efficiency.

According to its information, the equipment had the following characteristics:

1. Semi hermetic compressor with a 15-ton capacity.
2. Power supply 440 volts, 3 phases, 60 cycles.
3. Condenser of the shell-and-tube type.
4. Receiver equipped with a packing gland type king valve on its outlet.
5. Operates with R-22.
6. Equipped with a dual-pressure control type.
7. Liquid line was equipped with a liquid-moisture indicator.

According to its investigation:

a) The operating pressures were higher than normal.
b) The operating amperage (FLA) was high, but within the ID plate's value.

c) The temperatures of liquid and suction lines were higher than normal.

d) Liquid indicator (bubbles).

Questions

1. Do you think this change is a good idea?

2. Do you believe that in this case, glycol can serve as water replacement?

   If you think this replacement is feasible,

   a) What other changes may be necessary?
   b) What may be the temperature difference between the two condensers' supply and return lines?

3. What basic characteristic in this case differentiates both substances?

4. What do you think is (or are) the solution (or solutions) to this problem?

==

EXPERIENCE 103-A

It was the beginning of summer, and a lady director of a group of offices located in a building on Second Avenue and 118th Street approached the school for help. She asked for help verifying the air-conditioning supply to the offices at her charge.

According to this lady, during the cold season—autumn, winter, and spring, the building supplied the heating service; via radiating units (hot-water convectors of the fin-tube type); distributed over the place. For the air-conditioning supply, the place was equipped (a couple of years before) with a 5-ton-capacity air-cooled A/C system.

We asked the lady for some printings or plans for the A/C system. She didn't know of the existence of such.

The superintendent of the building had nothing to do with the A/C equipment's operation.

At the lady's request, he tried to put the system in operation, but he couldn't find the reason why it's not operational.

With a couple of students, we went to the mentioned place to verify the problem.

The A/C system had the following characteristics:

1. A semi hermetic unit, using 440 volts, three phases, 60 hertz.
2. Air-cooled condenser (fan motor working at 230 volts, 1 phase, 60 Hz).
3. A step-down transformer (440–115 volts.)
4. A cooling thermostat (115 volts)

5. A dual-pressure control (high and low)
6. A liquid line solenoid valve (115 volts, pump-down system)
7. A timer (115 volts)
8. R-22 refrigerant

When installing the pressure analyzer, it was noticed that the compressor's temperature was kind of warm, as it has recently been in operation.

In addition, the following were noticed:

1. Pressure-control (high and low) adjustments:

   Low-pressure control Cut-out point = 45 psig (22°F.)
   Differential = 20 psig (16°F.)

   High-pressure control; Cut-out point = 300 psig (130°F.)
   Cut-in point = Automatic reset

2. The system pressures (R-22) were found as follows:

   Suction pressure = 50 psig
   Discharge pressure = at room temperature = 55°F (92 psig)

3. A temperature sensor installed outside, to "feel" the ambient temperature, was found set to "open," disconnecting the liquid line solenoid valve and (through the timer) the condenser fan motor's contactor at a set point of 40°F or 20°F differential (cut-in point of 60°F).

Note: The outside temperature today was raiding on the upper 50th.

When checking the electrical connections, it was found that one line from the transformer {1} came directly to the condenser fan motor contactor's coil (5) and the thermostat (2). The thermostat outlet went directly to the liquid line solenoid valve (3).

The other transformer's line was found to be connected in a series circuit, with the outside temperature sensor (6) then connected simultaneously to the liquid line solenoid valve (3) and to the timer (7). From the timer {7}, to the condensers fan motor's contactor coil (5).

Two more lines came out from the timer, to control other electrical components, i.e., the compressor and evaporator's fan motor.

Note: See diagram below.

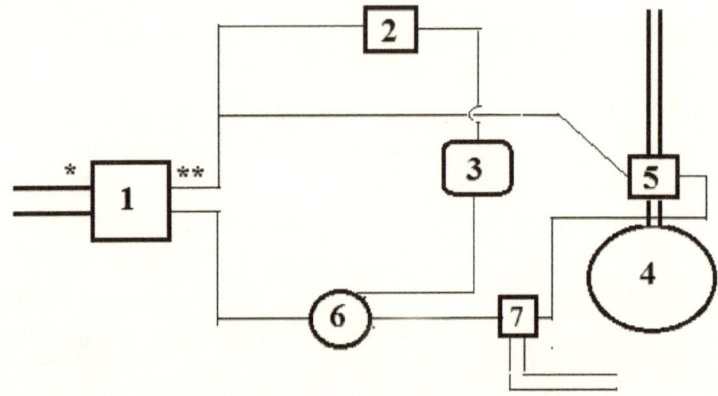

INDEX.

1. Transformer * 440 - ** 115 volts.   2. Thermostat 115 volts.
3. Liquid line solenoid valve 115 volts.   4. Condenser fan motor 230 volts.
5. Condenser fan motor contactor.   6. Temperature sensor (outside)
7. Timer 115 volts.

Questions

1. Is the thermostat directly affecting the compressor's operation?

2. Why does neither of the compressor, the condenser, nor the evaporator motor work?

3. What procedures should we perform to find the device affecting the three motors' operation?

4. What should we suggest doing to allow the A/C system's motors to work?

===

EXPERIENCE 104-A

This service call was to a supermarket. The client complains that for the last couple of weeks, the products stored in this commercial refrigeration fixture have been damaged too fast. The required +/-40°F was used as a cold-cuts display cabinet for products such as cheeses, hams, salami, mozzarella, and other products that, according to what the customer said, get moldy. And because of this problem, these could no longer be used or sold.

The equipment had the following characteristics:

1. Semi hermetic unit of 3/4 hp condensation by air, using a thermostatic expansion valve.
2. It worked with 115 volts, 60 cycles, 1 phase.
3. Used R-134a.
4. Filter-dryer and liquid indicators in the liquid line.
5. Solenoid valve on the liquid line.
6. The evaporator was located in the upper part of the cooling compartment. Air movement was by natural convection.

When connecting the manifold, although everything was normal, the operating pressures (discharge and suction) were a bit lower.

The liquid indicator, at the liquid line, showed "Full," the liquid line temperature a little low. But all the other operating conditions were almost normal. The suction-line temperature was a bit lower than normal.

By observing the storage compartment, it was noticed that the grills where the products were kept were wrapped up with aluminum foil leaves. When I asked the client why of these leaves, he said that it was for preventing the grilles' metal from getting oxidized or rusted.

Questions

1. Do you have an idea where the problem might be coming from?

2. Why are the operating pressures low?

3. What should you do or recommend to fix the problem?

4. What explanation should we give to the customer?

5. Why was the temperature of the suction-line lower than normal?

EXPERIENCE 105-A

The service call was from a delicatessen store, and according to the customer, for the last couple of weeks and after a repair (first service company made some changes), two refrigeration service companies had worked in the self-contained beverage cooler without an finding explanation for the device's poor cooling effect.

Required temperature = 40°F.

The condensing unit was located at one end of the cabinet and it was working.

The equipment had the following characteristics:

1. A semi hermetic unit of 1/2 hp.
2. Condensation by air.
3. Used a liquid control of a thermostatic valve type.
4. It worked with 115 volts, 60 cycles, 1 phase.
5. Used R-12.
6. Operation controlled by a thermostat.

Temperature at the condenser's location: 80°F.

While talking to the customer, I stood at the beverage cooler's condenser side (front of the equipment). I inclined it, and by using my hand, noticed something strange in the airflow through the device's condenser. To make myself sure, I used a sheet of paper, put it in front of it, and found that my suspicion was right; the fan was moving the air in the opposite (wrong) direction. Meaning it was pushing the air out, instead of sucking in.

Questions

1. Do you think, the wrong rotation of the condenser's fan motor can be the reason for the poor cooling effect in this equipment?

2. What other tests should we perform?

3. What should we recommend doing?

4. To solve this problem, can we invert the actual condenser fan motor blades?

5. If the condenser fan motor had to be changed, what rotation characteristic should the new motor have?.

EXPERIENCE 106-A

It was two in the afternoon. Probably due to the proximity to where I lived, the dispatcher sent me to this service in a restaurant located in New Rochelle. The person in charge informed me that for some time one of the ice-making machines failed to make its work. The person also told me that there have already been several mechanics from other companies, and they could not find the problem.

The machine was switched off. For me, this machine does not appear too old. Before putting it to work, I verified it electrically and did not find any impediment to energize it.

At this machine location, the room temperature was approximately 80°F.

The system had the following characteristics:

1. It had a semi hermetic compressor of 3/4 of a ton.
2. It worked with 115 volts, 60 cycles, 1 phase.
3. It used refrigerant 12.
4. Condensation by water (Municipal) Tube within tube condenser type.
5. Defrost by hot gas.
6. Equipped with a dual (high and low) pressure control.
7. It had a thermostatic expansion valve.
8. The evaporator (cube molder, upside down) was located so that the ice cubes were formed on the bottom.

To make the ice, the water pump delivered the water by means of an upward "spray," in a back-and-forth movement (of approximately 30 degrees angle). During the system operation, the ice would form inside the mold-cube spaces.

One of the cubes served as a "guide-sensor," and when all the cubes were about 80% formed, the master-cube activated the timer, which in turn (about 5 minutes later) energized the defrosting solenoid valve, which allows the hot gas pass to the evaporator, so the formed ice cubes were harvested. I check the sensor, the low-pressure control, and the timer settings; all of them seemed to me set too low.

The manifold's hoses were connected to the equipment's suction and discharge pressure valve's connections. The machine was set in operation, and at the start and during it, the operating pressures and temperatures were checked.

At about ten minutes operating time:

1. Suction line temperature (6°F.) seemed very low.
2. Liquid line temperature (90°F.) was normal.
3. Condenser's water temperatures: in = 50°F out = 60°F. (normal).

Near the end of the cooling cycle, the suction pressure seemed too low at 6 psig (-6°F). I assumed that probably by customer demand, some service mechanic made the adjustment (sensor, low-pressure control, or timer) lower to get the ice harder.

Operating cooling cycle was timed at approximately 22 minutes (it seemed a little too long).

The machine went into defrosting, but after 15 minutes into the harvesting stage, the ice did not fall in the amount it should. Besides, some of the cubes that were produced appeared smaller than normal.

Questions

1. Initially, what type of electric test should we conduct to make sure that the equipment will be safely energized?

2. Without further investigation, can we diagnose the problem in this machine?

3. What test and what tools can we use to decipher this problem?

4. What should be suggested to repair this ice-cube maker?

5. What should be the normal operating pressures for this system?

==

EXPERIENCE 107-A

This former student called me to get a consult or to ask me for some aid. He was called to verify the operation of an air-conditioning system in a Chinese restaurant located at Roosevelt Avenue, Queens.

According to what he told me, the customer said that since a company made a job (changed the fan motor in the cooling tower), the A/C system was no longer cooling as before. After the repair, the company's mechanics returned a couple of times, but the problem continued. Seems, the owner has had a problem with the service company, and they did not return.

The A/C system had the following characteristics:

Condensing unit, located in the basement.

1. Semi hermetic unit, capacity 25 tons.
2. Working with 440 volts, three phases and 60 Hz.
3. Used R-22.
4. Water-cooled condenser shell-and-tube type.
5. Water supplied by a cooling tower located on the roof.
6. Water circulating pump located, at the cooling tower's base side.
7. Equipped with High and Low pressure controls.

Normally, water from the condenser returns to the top of the cooling tower's sprays; then falls through it.

Since the problem was in the cooling tower, we went to the roof.

It was hard for me, because I had to climb a ladder. But, as I got to the cooling tower; I was able to check its features:

a) The water pump, used a three phase motor 440 volts, 3 phases and 60 cycles.

b) The electric fan motor (440 volts, three phases and 60 Hz.), was mounted on one side, top of the cooling tower.

c) Through the set pulley-flywheel and two belts, the motor drove the blades (mounted on top) moving air through the tower.

The set was not in operation.

Check the electric fan motor, which appears brand-new and did not present any defective electrical o mechanical conditions.

The set, flywheel, pulley, and belts was verified, and they looked well.

So far, everything looks fine.

Questions

1. After put it in the cooling tower into operation, what should be the next step?

2. What do you suspect the problem was here?

3. What should be the rule for airflow through the cooling tower in this or any other case?

4. If necessary, what can you do to change the rotation of the motor or blades?

5. Knowing what you are doing, how long should it take to notice the problem here?

==

EXPERIENCE 108-A

This service call was to an Italian restaurant located on Second Avenue. The person with whom I spoke told me that one of the ice-making machines, for the past couple of weeks, had some operational problems that interfere with the ice production.

This person also rendered the following information: If the machine were disconnected for a few hours, then put to work, it produced ice, but then in the next cycle, it stops the production again.

It led me to where the machine was located. The equipment was de-energized.

As the person told me that the problem appeared when the machine was in operation, I decided to omit verifying the equipment's electrical condition.

The ice-maker device had the following features:

1. It was a semi hermetic unit of 1/4 hp. It worked with 115 volts, 1 phase, and 60 cycles.
2. The condensing unit was located at the top of the ice storage compartment (bin), and thus above the evaporator.
3. Used a water-cooled coil-within-a-coil type condenser (municipal supply)

   The water flow to the machine was controlled by a solenoid valve.

4. Worked with R -12.
5. To defrost-harvest, hot water was used.

For the defrosting purpose (5), the machine was equipped with a water tank, located above the bin, next to the compressor. The compressor's discharge line passed, first through the coil, inside the water container, then, through the condenser.

Heat from the super-heated vapor refrigerant in the discharge line, at an approximate temperature (I assumed) of 175°F. or so; heated the water. At the bottom of this water container, there was a piece of tubing with a solenoid valve mounted on it; at harvesting time, when this solenoid valve was energized, it let the hot water poured; on top of the evaporator. A timer controlled this solenoid valve operation.

The water to make the ice, was circulated by a pump; this water went to a spreader, that supply it the water at the bottom of the evaporator in an upward back-and-forth motion of about 30° angle and the (water) that was not changed into ice, it returned to the tank; to be re-circulated.

As mentioned above, the water flow to both equipment tanks; was controlled by a solenoid valve, affected by a timer. Both water tanks; for the water level control, were equipped with floats. And, at this time, they both were filled with water.

The water supply lines to the two tanks; were provided with separate water filters.

The machine was put into operation. It worked perfectly; the water in the preheat tank reached a defrosting temperature of approximately 170°F.

Questions

1. At this moment, can you suspect what the problem may be? If not, let's put the machine in operation.

Operation

The ice making cycle took about 15 minutes. At this time, the timer stopped the compressor; and energizes the solenoid valve; located under the hot water tank. Hot water was poured over the evaporator, and the ice began to harvest falling into the bin. After about three minutes (harvesting time), the compressor began to work, and the defrosting solenoid valve was de-energized.

But something is wrong: The water in the deposit for defrost, is very scarce.

----

1. Do you have an idea, about what the problem may be?

2. What do you do, to find out about this water problem?

===

214

EXPERIENCE 109-A

All the requirements that were needed for the Technical Trade School; preparation of the place and the rest of transacts were performed simultaneously with the attendance to and the study of the "A.S. Degree in Applied Science" (Environmental Technology) at Technical college.

After I visited the buildings department, the next step was to find an architect so that he represented me, before them. With my studies completed in college, I elaborated (a 1' = 1" scale) plan of the school floor by floor. So when I found the architect, I showed him those plans, and that helped me; so the representing cost was not so expensive. This man, visited the school, and among other things he asked; for the installation of a water sprinklers system (fireproof) in each floor. For this effect, I went to a company that was near the school and arranged such work. It cost me, a little but $2,000. Also, it was necessary to change the way the school's street door, opened.

One thing follows the other. Arrange a premises inspection, with the fire department. With them, there was no problem; a ladder car came, with ten firemen, including a captain.

While I spoke with him, the rest of men inspected the place from top to bottom, and they reported everything to the captain. Once they finished the visit, the captain told me, that everything was well; that they would submit the report to the education department.

Then, I went to the buildings department, to ask for the inspection. They gave me a date. In the indicated day, an inspector arrived. Between the things that he found it, was the necessity of two devices (an air supply fan and an air extractor) that would have to be installed in the back part of the first floor (shop) (the second floor was about 4 meters shorter than the first; so, the first floor had a small roof), in that space; would had to be installed both motors that would fulfill those requirements. I had no idea, of where to go to obtain those motors.

The currency was depleting and still, I did not see a clear horizon. On this day, I went to No name (a warehouse for refrigeration and A/ Conditioning products), to buy materials for practice in the school; I saw on the floor a fan of the mushroom type of about 2.5' of diameter, which may be installed in the first floor's roof to move the air in.

Ask the employee and I confirm myself, it was a ventilator. By curiosity, asked him if they also had the extractor? He answered, no. I did not wait for much more. I bought one of the devices.

Questions

1. Since I had the ventilator motor, do you have any suggestions for the extractor motor?

2. Is there anything that can be done to fulfill this requirement, without expending much money?

3. What should we do?

==

EXPERIENCE 110-A

By this time, I had this former student helping me in the school as an instructor. He had already finished his Environmental Control studies at Technical College. He had married with his countrywoman of the name ——. She knew my wife and me. She was a story aside. Once, I had to call him to their home. I asked for Mr. ——. She "corrected" me and said, "You mean the engineer ——. Right?"

I do not remember well, but after a while, Mr. —— stopped working in the school. Sometime later he called me, so I could advise him in a work that he was going to carry out in a supermarket of the zone.

I went with him to the place, and I recommended to him to make the job's price, including the materials; that way, he could manage better the gains. Instead, he let the client convince him, and he was only able to charge for his work.

He expended a long time doing that work; I thought due to that, he brought a friend to help him. Specifically, his friend was the person who performed the entire piping connections soldering job.

I had to go to the place several times to help him in solving several problems he was facing. One of those was a problem with the rigid tubing connections (liquid line ¾" and Suction line 1 1/8") that ran about 50' from the compressor to the evaporator. They were using rigid tubing (lengths of 10' each).

After they finished the entire soldering job, they were checking for leaks; they pressurized from one end, but they did not have any pressure signal at the other end of the testing line.

Questions

1. What do you suspect was the problem here?

2. Any suggestions to prevent these problems?

3. What do you think could be done during the entire process?

===

EXPERIENCE 111-A

With the purpose of cooperating with the A/C service company's school owner, I suggested elaborating a plan to work together with the supervisors. This plan involved the preparation of some courses for the company's personnel.

The owner got me in contact with his son, as well as one of the supervisors. My suggestion was to offer appropriate and separate courses for each group of workers: apprentices, helpers, mechanics of third, second, first class, etc. The plan started up. But the supervisor didn't have the slightest understanding of what my idea was; one day, he appeared with a number of workers, and he tried to teach a class of general interest for all. It did not work.

This man didn't have the slightest idea of teaching.

In this case it was not even appropriate; he brought together a group of workers and tried to teach them a topic of general interest. Why? The participants had different knowledge levels, experiences, etc. Instead I gave the supervisor a list of the topics that can be covered; but this guy, disposed most of the topics, among them the electric ones referring to measuring instruments, alleging that they were not necessary, and putting all types of inconveniences that for me did not make any sense. For him, the electrical instruments didn't have the slightest interest in troubleshooting the equipment. So you tell me.

He preferred to put to work its own ideas, without counting much with mine. One day, he came before the workers, and "he prepared" the equipment, with faults of its own invention; a little later came several mechanics to diagnose the faults that the supervisor had prepared.

Without further information, he put the mechanics to find the problems created by him.

For about forty-five minutes the supervisor gave "instructions." The result! Two of the compressors of the equipment that had been donated to the school's classes were burned out. The two damaged devices had the following electrical characteristics:

1. One and a half (1 1/2) tons, 230 volts, single phase and 60 Hz.
2. Three tons, 440 volts, three phases and 60 Hz.

Nothing good came out of the shining supervisor's ideas.

The supervisor offered to obtain with the manufacturers the spare part for the burned compressors. He took some information (I didn't know what) from the compressor's nameplates and said that that should be sufficient to obtain the compressor's replacements. He wasn't even able to obtain the spare right parts for both of the burned-out compressors. They sent (twice) the wrong spare parts.

Since the two compressors came from different sources, I decided to speak directly with the representatives of the companies selling both compressors.

One last word about the supervisor. I think he belonged to the group of people in the company that was not in favor of the school.

Questions

1. What should we do to troubleshoot the electrical installation in an A/C system?

2. If the system it is not working, what should be the first test to carry out?

3. What and how the electrical instrument should we use?

4. If the system is energized when using a voltmeter to verify the opened condition of a switch, fuse, circuit breaker, overload, or others. What reading should the meter register?

5. Same question as in # 4. But, system de-energized and using the ohmmeter?

6. What other information should be taken, to make sure in getting the exact replacement?

==

EXPERIENCE 112-A

This person particularly caught my attention. He came to the school, to register itself in a Domestic refrigeration course (by this time, the school did not have yet, any entrance examination requirements); he paid his registration and began the classes Monday through Friday (nights from 7:00 to 10:00 p.m.).

He asked me permission to record (in cassettes) all the classes (I did not put any objection) and to facilitate the recording, he seated on the front row near the instructor's desk.

The first week passed; then of course, the first test came. I distributed the examinations papers, and I gave the corresponding instructions to answer it. Everyone started working in the test. I noticed that the student; did not write anything. I let pass a few minutes, and as he continued in the same attitude, I called him to the desk and I asked him what happened. He answered that he didn't know either how to read or to write. However, he told me that he could make the examination; by oral form. I proceed and in a low voice, I asked the questions and I was impressed, he was able to answer each one of the questions perfectly. He told me that I did not have to worry about him, because he knew what he was doing.

At the end of the domestic course, he subscribed in a washing/drying clothes machines course. He confided me in addition, that he had a business, where he sold newspapers, magazines, cigarettes, etc. and once, he learned enough (when finishing this course), he would open his own service business. Thus, he did it, at the end of the course; he traveled to New Jersey, bought a pick-up type truck and opened his business.

As he told me, he quitted the articles' sales business that he had. All the mornings, at about 4:00 a.m., he went through Manhattan' streets; near to where he lived, and he gathered whichever refrigerator, air conditioner, washing of clothes machines, etc. that he found; took them to his business place. He repaired-painted them (if necessary) them, later he sold them.

In addition, he was able to get some clients; in the commercial refrigeration branch. According to what he told me, he repaired many of the equipment, following the information he recorded in the classroom and shop.

It seems to me that a client satisfied with his work; requested from him the installation of refrigeration equipment for cooling beverages and beer; in an empty walking-box that he had in his business' premises.

He visited another one of its clients, in another store and as he could, gathered the refrigeration system information (similar to which he was going to install).

He adjusted the price of the work ($7,000), he requested 50%, bought the materials and in the length time of one and a half days, he installed and put the equipment to work.

The satisfied client, not only paid him the rest of the money but in addition, he gave him a gift of $1,000 more. After finished the work, the client confided to him; that one of the companies (the cheapest one) had requested $12,000 to make the job.

He came to the school (in several occasions) and he told us of its life. He was a barbarian; he sat on the classroom's floor, he recommended the students; to pay a lot of attention. He told them, that what he was, had been and have, because what he learned in that school.

Once, I had a problem with my refrigerator at home. I called this man, and he told me: "Mr. Jimenez, do not worry. Just tell me how to get there." He requested a second, then he asked me for the directions to get to my residence; which I gave him the clearly way possible.

He told me "In about 45 minutes I'll be there." I went down and I wait for him outside, in front of the building in which I resided. As he told me, it thus was. He was there, in a little more than a half an hour. He brought a brand new refrigerator. It caused me curiosity that he arrived without any problem; I ask him how he did it? He said to me, "Very simple, I recorded the indications that you gave me and I followed them."

I had always thought, "What shall this man do if had known to read and to write!"

Regardless of who he was, what it is your opinion about this guy?

==

EXPERIENCE 113-A

Given a hand to (a former student), he bought parts and components in an R and A/C supply store; located at 125th street between 2nd. and 3rd. avenues in Manhattan. In one of those visits to the place, the owner asked me if I had heard of the RSES group (Refrigeration Service Engineers Society) and I answered to him, that I belonged to that organization.

The next time we were there, the man asked me if I would like to teach; for the RSES organization. He told me, that he had spoken with the president of the New York chapter and if I liked, he would call me to complete the details. I answered to him! Yes, that I would do it.

A couple of days later, the president of the RSES Chapter in New York. (In those days Mr. Stanley Hollander) got in touch with me and we arrived at the agreement that I would teach the courses that that organization offered to its affiliates. He emphasized, in getting my commitment.

Thus, I began this relationship with RSES. By a few months, I worked in a public school in Corona, Queens; starting at the beginning of the autumn and ending at the end of spring; Mondays thru Fridays, evenings, from 07.00 to 10:00 p.m. Later on, I continued in the Queens Community College. With the same RSES organization, in an adult "Continuing Education" program.

Once I had the RSES education books (made by groups of people relating to the organization), I felt satisfied with my knowledge, I did not had much to prepare myself for the classes, because in the books prepared by me for the Colombian Navy (and for my school), were very much as completed as theirs.

===

EXPERIENCE 114-A

I believe it was around this time that Ricardo (my son) had a business relationship; with these people in a place of Connecticut, distant from New York about 25 miles. They, distributed, sold, maintained, etc. all the related links with decorative fish tanks; of all sizes.

Ricardo got into contact with them, because he had in his home a varied and beautiful decorative fish-tank, with a size of 8' x 4' (32 cubic feet) and they supplied him, of all the necessary implements to maintain it.

I also was involved with the hobby, and I had in my apartment a much smaller fish-tank but with a good variety of tropical fish. By his relationship with these people, Ricardo began to give refrigeration service to their equipment. I think, he had a good business relationship with them, I don't know why or when, but some time later; he discontinued that work.

Among other experiences, I remember that Ricardo commented to me of a work-repair, make it by some service company in their premises before him: "Incredible" but truth.

Some time back, the service company installed in the warehouse, a 15 tons air conditioning system (R-22). We thought that the customer asked, for the equipment's installation, inside the premises. So, they installed the compressor, of the air-cooled system inside; at about 3' away from the wall of the machine's room; the condenser was located, near the rooms outside wall.

The discharge line from the compressor, ran horizontally and directed on the wall's direction; when arriving at it, they threw the pipe vertically making use of an elbow (forming a 90° angle); to be able to get it through a window; located at a height of about 5 feet of the floor; then, another 90° elbow and so on. On the outer part, they finished the connection.

They completed the job, by charging the refrigerant to the equipment; they started it and let it running.

Questions

1. Can we see anything wrong in this job?

2. Can you forth see, any future problem with this installation?

3. Could you have, a different idea for this installation?

EXPERIENCE 115-A

On this day, I was driving up by the Third Avenue, northbound, when suddenly somebody in another truck behind me began to toot the horn with as much insistence that I slow down to the right, and with the left arm, I made a signal to indicate to him that he could pass.

Nevertheless, he continued with the same horn toot. I decided to stop my vehicle.

Then the truck stopped behind me, and the driver got off and walked toward me. Then I recognized him; he was one of my National Skills Center former students. He was a Cuban, his name Jaime Mass.

He greeted me very enthusiastically, and once on my side, he told me that always he had remembered me, that he has been eternally thankful for what he had learned with me. He told me that he was working as a mechanic in this company.

He confided to me that to enter this company, they asked him what experience he had; he told me that he had lied and told them that he had worked in Cuba. They performed an examination on him, and they hired him.

Performing his duties, he had a lot of problems, but by using his own time (often at nights, he returned to the jobs to make adjustments, and make sure that he did not have any problems with his work or with the clients). Up to this point, everything was OK. He told me that many times he stays till late hours at night, reviewing the notes that had taken in the school.

When finishing the course, he dismissed and thanked me very much. Years later, I found out, that he had traveled; he was working in Miami, Florida, on his own business.

Questions

1. What do you think of this guy' behavior?

2. Would you do the same things that he did to keep his job?

3. Returning at night to verify the system's operation?

===

EXPERIENCE 116-A

It was Saturday morning; the problem appeared on Friday night. I was sent to this super-market, located in 9th Avenue and 56th Street. The person in charge directed me to the business's cellar. The place was flooded, and a lot of the stored merchandise, had been damaged. Several employees were cleaning and removing the water, as well as the damaged merchandise.

The flood was caused by the faulty operation of the water pump, whose work was to remove the water from the refrigeration machine's water-cooled condensers and the condensate water from the A/C equipment.

The company sent a mechanic, who came Friday afternoon, took all the water-pump's information, got a new pump and change it, it put it to work and it went away.

Questions

1. Can we say if it is anything missing in this procedure?

2. What should we do after the installation of a new water pump?

3. What must we do before leaving a performed job?

===

EXPERIENCE 117-A

It was spring, and a former student (George) called to consult a problem that he had with an air-conditioning system, window unit type, that he was repairing in his shop, located in a basement.

The device had the following characteristics:

1. Capacity of 24.000 Btu (2 tons)
2. Electric features: 230 volts, single phase and 60 cycles.
3. It used refrigerant 22

He told me that he had changed the burned-out compressor.

According to him, he carried out all the recommended procedures:

1. Checked for leaks.
2. Pulled a deep vacuum.
3. Charged refrigerant by weight, and calculated the operating suction and discharge pressures;
4. The applied voltage value was verified, as well as the current consumption: Start (LRA), running (FLA).

Room temperature = 80°F.

At my arrival, the system was completely assembled. George told me, that in this place for some reason, the A/C equipment takes very long time (more than usual) to start cooling. He had experienced that problem, with different A/C equipment that he had tried here.

We started the equipment operation, and found the following characteristics:

a) The operating pressures were a little bit too high:

Discharge 250 psig. Suction 75 psig.

b) Higher than normal, running amperage (Compared w/system's Data plate).
c) Higher temperatures:

Suction line 65°F. Liquid line approximately 95°F.

d) Air temperature difference through the evaporator 10°F (In=80, Out=70°F)
e) Air temperature difference through condenser 40°F. (In=80, Out=approximately 120°F)

By observing the system's operation (page 1 in this experience), and according to the pressures and temperatures obtained . . .

Questions

1. What do you think may be the problem in this equipment?

2. Do you have any objections to the job performed here?

3. What other information is needed to get the proper operation?

4. What else do you think can be done?

==

EXPERIENCE 118-A

By 1984, I was trying to obtain an environmental certificate (Pollution Control) (not that I need it, but because I always like to be and to have certifications that served me to make my teaching work better); I even called the Environmental Protection Agency in Washington, without any positive result.

One of those days, came to the school a former student; among other things he told me that he had obtained the certificate that I was looking for. He gave me the address of a public school, at (Bruckner Boulevard) in the Bronx; where they prepared people for that examination.

An afternoon I went to the cited school, at the hour that he indicated me (classes took place Mondays thru Fridays at evenings, between 6 and 10:00 p.m.). I met with the instructor, and we were speaking for half an hour about the course's preparation content; until the students start to arrive.

According to our conversation, the instructor reached the conclusion that the preparation I had could be enough, and I didn't have to attend the classes.

In an agreement with that exchange of ideas, I thought, that the information that I had in my head; from the work in the railroads and the trade school, studying times in Colombia (now, about more or less 40 years). He gave me the next examination's date (within 6 weeks), and I had to get back at the indicated day, a Friday.

On the cited date, I went to the school; and presented myself to the instructor. It was 5:30 p.m. He was surprised of my punctuality, and he asked me if I was ready; I responded to him affirmatively; then, he gave me the examination pamphlet; the test consisted of 100 multiple choice questions. I seated, and I began to answer the test.

As I had anticipated, I did use of the long time ago acquired knowledge, stored in my brain; some questions were common sense. A little before 6:00 p.m., the students began to arrive. I answered without great problems 98, of the 100 questions; but, I got "stuck" in two of them. I was trying to find the answers for those two questions but, I answered the best I could; and gave the examination sheets back to the instructor. About 35 or 40 minutes had passed.

The instructor asked me, if I had already finished? I said yes. He was surprised once again. He gave me his business card, and he told me to called him the following Monday to give me the results.

Throughout the reading of this experience, do you think that we can remember things that had happened more than forty years ago? And with them, will we be able to answer a test?

EXPERIENCE 119-A

With this mechanic, I had several experiences; it seemed to me, that the other mechanics in the company did not have a very good reference about him, because on the following day when they asked me with whom I had worked the previous day (something that happened frequently) when giving them its name or responding to them with whom, they always showed an expression like meaning (I knew it).

Anyway, this time the dispatcher sent me to a coffee/restaurant located at Lexington Avenue and 41st Street to help the aforesaid mechanic, who was in the same place by several days. As always it happened to me. When I was sent with that intention, he received me with the usual question, "What are you doing here?" This situation was not the first with this mechanic, and of course, my usual answer was "The dispatcher sent me to help you!"

In any case, I asked him, what seemed to be the problem, and he said, "The heating, electrical system does not work!"

At my question, he indicated that the equipment was located in the ceiling, above the zone that I was working for.

Questions

1. What other question we may ask to the mechanic?

2. What should we do to locate the problem?

3. In the case that printings for this equipment were missing, what should we do?

4. Should we talk to the customer?

===

EXPERIENCE 120-A

While I had Technical Trade School, on Second Avenue between 117 and 118 Streets in Manhattan, New York City, I received a visit from this man who came specifically from Las Vegas, Nevada.

He was interested in the course of air conditioning. He knew some person from Los Angeles who told him, about my school. We fixed the cost of the course.

During the training, he came to the classes accompanied by his wife. We had classes of 3 hours per day Monday thru Friday (2:00 to 5:00 p.m.) by a whole month for a total of 60 hours.

According to what he said, he had worked in the casinos by sometime, helping the people who take care of the A/C equipment; of several of them.

He showed some practical knowledge; I believe that by the acquired experience, it was not so difficult for him, to understand the work, maintenance and service to this type of systems. He came here, to acquire the A/C's technical-theoretical part.

In the end, he passed the course final examinations. He said that he was very satisfied and he thought he had done very well, by coming to take the course with me.

===

EXPERIENCE 121-A

For several days, I got the feeling that my partner was experiencing some problem. I did not want to get involved in his personal business; but out of curiosity, I decided to ask him the reason for his worries.

He told me that a 25-ton air-conditioning equipment that he was installing did not attain the system's operation.

He told me that he had designed the printing plans for the electrical system; that he had followed them, but it did not find the problem. He "had verified" the equipment's both circuits; the high voltage, 440 volts power supply (fuses, circuit breakers, installation, loads, etc.) as well as the low-voltage circuit of 24 volts at the transformer's outlet, all controls, and it was not able to get the equipment to start.

He claimed that both the equipment electrical installation as the electrical printing plans were identical.

I offered myself to help him, but to start, I asked him to bring the printings. So I could take a look on them.

The following day, he brought the set of printings.

Questions

1. How many electrical circuits an equipment like this should have?

2. What are the electrical components in each circuit?

3. By using the printing plans. What do we think, should be the right procedure to follow in finding this problem?

4. On the actual equipment; what kind of tests should we perform and what kind of instruments should we use?

    a) On the electric circuit's loads?

    b) On the electric circuit's controls?

===

EXPERIENCE 122-A

This operational problem, often happens especially in Air Conditioning equipment used in commercial establishments; warehouses, bodegas, Delicatessens, etc. where the equipment most of the times; is placed on the top of the front door of the establishment.

These type of equipment has, generally the following characteristics:

1. Voltages of 230 volts, single phase and 60 cycles.
2. As all the A/C set is located in the same place; an air-cooled condenser is used.
3. Their cooling capacities are, usually between 18,000 and 48,000 Btu (1 1/2 to 4 tons).

For the A/C equipment to achieve the cooling effect in the conditioned space, it must remove simultaneously (or at a time) the air humidity content as well as its temperature, to the desired points. In the summer, from ± 80–90°F (± 27–32°C) (RH ± 20-60%) (Dew point from 24–60°F) to ± 64–72°F (± 18 - 22°C) (RH ± 60%).

Most of the times, the air conditioner equipment; works for spaces (constructions) that are not conditioned (not insulated). Besides, the access doors are in constant movement (open-closing) and the entrance of the outer air and therefore the humidity; are constantly getting into the place. This dripping water, often presents inconveniences for the customers moving in and out.

So, for an A/C system, to start "cooling" the space; it must remove from it, simultaneously, the excess of humidity (latent heat); and the cooling heat (sensible heat).

This way, the "production" of water is constant and un-evitable. Although the equipment in normal conditions, can "use this water"; and to practically return it to the outside space; in most of the times the dripping water, is almost impossible to avoid.

Questions

1. What can we do to minimize or eliminate this dripping water? ! Without affecting the A/C system operation!

2. Do we have an idea of how much water is produced, under the above conditions, in a one-ton A/C system?

3. How is the water produced by the equipment used by it?

4. What are some "mechanisms," used by the A/C equipment to use that water?

===

EXPERIENCE 123-A

For this job, this morning I was called to the office; one of the supervisors wanted to talk to me. I understood that this man had some friendship; with the people at today's job place. A bar located on Second Avenue; and on the forties Streets.

He asked me, if I could be able to remake an electrical installation; in an A/C equipment of three (3) tons capacity. I ask him, if there were an electrical printing plan that I could use, and he said yes.

He also told me, that any material or anything else that I needed; they, at the bar, would provide it.

I went to the place, at my arrival, at about 9:00 a.m. they, who already knew about the arrangement; greeted me and took me to the equipment location on the back of the bar.

They explained to me, the cause of the problem: A small fire in the place had taken place; some of the electrical installation to the A/C equipment was damaged, and because the A/C unit installation it was kind of old; they decided to remake it. The A/C equipment had no other problems.

They said to me that, the supervisor had already verified the components of the equipment and that, according to, were well.

I told them, that until I verified all the electrical components, by myself; I could not carry out the work. I didn't know, but seemed to me that they knew how I worked; they agreed and said that, that was very well.

This A/C equipment had the following characteristics:

1. Sealed unit.
2. The equipment used 230 volts, 3 phases and 60 Hertz.
3. Condensation by water (cooling tower)
4. Condenser "shell-and-coil" type.
5. The equipment worked with R-22.
6. Timer. This device used 230 volts.
7. Step-down transformer 230 volts to 24 volts.
8. Controls: selector switch, contactor, thermostat, dual high-low pressure control.
9. Thermal overload, etc.

Questions

1. Should we be able, to remake the installation of a system likes this?

2. What type and wire size, should we use in the high voltage part of the electrical circuit?

3. What type and wire size, should be used in the low voltage part of the electrical circuit?

4. What other important items we may observe, during the wiring of this job?

==

EXPERIENCE 124-A

While working with HVAC Tech Inc. School in Brooklyn, in one of the A/C classes I had a student who worked in a company in New Jersey. According to him, the main occupation of this company was investigation.

He told us that in one of the experiments that they carried out, a refrigeration system was used, to study metals reaction to exposure to different temperatures: This equipment was designed so it could lower the temperature down to 200°F (129°C) below zero.

Sensors were attached to the metal object of the study, with the purpose of measuring its reactions to variables such as hardness, flexibility, expansion, rupture, brittleness, contraction, etc. in response to the temperature variations during the entire cooling process.

This information awakened our curiosity, and when the A/C course arrived at the practice time, I asked the student to talk and ask the New Jersey company manager permission for us, the students and I, to assist there for a practice.

The permission was obtained. They assigned to us one of the technicians. When we arrived there, the technician was ready. He began by asking if some of us had heard of a problem in which one of the branches of the armed forces of the U.S. was involved in some process, because they had paid several hundreds of dollars in buying a hammer. I remembered that I had read something about it, and in fact, the price of this device seemed to me scandalous.

The technician said, *"As of today, you are going to understand why of that cost, and the reason for the so, high price."* In addition, he explained to us, that steel being one of hardest/strongest metals; is used for the elaboration of tools: Blades for plows grub hoes, shovels and other tools used to work the land in farms, and the hammer of our story; are used in climates of temperatures; extremely cold.

<u>He continued.</u> *"Here, it is where we enter. In this company, we study the alloys used in the design of those instruments-tools among our work, is to find the way to avoid the breaking and other difficulties that appear with those problems.*

The entire investigation as you may notice, it's very expensive and somebody had to pay for it."

He took us to the front of this, very well insulated box, with an exterior's size of about two cubic feet. It opened a door that gave access to its interior; in the inner space, of approximately a cubic foot; an evaporator and a great variety of cables with sensors attached to their ends; these cables were extended, through the walls, to outside of the box.

According to the technician, within the box, the metal to be studied was placed but, not before the sensors were connected/attached to it. The cables extended from the box, were connected to a board with many instruments: Thermometers ($°C.$), pressure gauges ($N \cdot m^{-2} \cdot kg \cdot m^{-1} \cdot s^{-2}$), etc. in what the different reactions from the material; within the box were registered.

A refrigeration system of the "Cascade" type, with two different types of refrigerants, was used.

This equipment consisted of two refrigeration systems; the one of the higher temperature (R-502); was a water cooled system; its evaporator, worked to remove the heat from the condenser of the second equipment; the low temperature one (R-23). The evaporator of this low temperature equipment was located inside the box; where the experiment was carried out.

After all the sensors were connected, the refrigeration equipment was put into operation. We observed the operating pressures, and as the thermometer was descending until arriving at; in about 30 minutes, at the desired temperature reading of 200°F (129°C) below zero.

Throughout the entire process, the technician was explaining and showing-comparing through the different instrument's readings; as well as the changes that the metal, object of the study, was experiencing.

This one was a very interesting experience for all of us. And in the end. for all of us; was very clear why of the high cost of the hammer.

We all were very thanked for the technician, as well as for the company, and we took a leave.

Questions

1. What should be your opinion, about this field trip?

2. Should you have asked any questions?

===

EXPERIENCE 125-A

For several months, I was experiencing problems; with my automobile, a Chevrolet Malibu.

1. The first time, I was returning home from the school. I was traveling through the Major Deegan Highway. I was in front of the Yankee Stadium. It was about 5 or 6:00 p.m. Suddenly, with no reason the car's engine stalled. Less badly, than I was near the exit, and somebody pushed the car and I was left outside the highway.

I called for help to an auto mechanic friend of mine; he answered me and said that as soon as he finished something that he was doing, it would come in my rescue. That could be in approximately 25 to 30 minutes.

In something as well as 20 minutes later my friend arrived; he opened the hood took a look there and it requested me that I give it the try and start the car. I gave ignition and without a problem, the car started. He reviewed it and it did not find anything strange. As the car was new, less of one year, he recommended taking it to the cars dealer.

Next day as soon as I had the opportunity, I took the car to the dealer. They reviewed it and after a while, they rendered the report. ! They could not find anything wrong!

2. It was 8:30 in the morning. Hour at what everyone goes, in a hurry. I was going to the school at 117th. Street and Second Avenue. I was driving thru Bruckner Boulevard; suddenly, the car's engine stalled again. This was the second time that this happened. The last time, happened several months ago.

Several people helped me in pushing the car to remove it from the road.

After a while, being there seated and remembering the last time that this happened, there had passed approximately 25 minutes, and I tried it and the car started without but.

Again, I took it to the car's dealer and explained the problem; they reviewed it and as the first time, nothing was found wrong.

3. The third time, I had an appointment with the NYS Education Dep. Commissioner at Albany. I had to be there at 9:00 in the morning. It was early, about 7:30 a.m. I went in company of my wife and my sister.

We stopped at the last gas station before arriving at our destiny.

After taking breakfast, we got ready to continue our trip. But, again, the car did not start. As the gas station, had a service mechanic, I requested to him that please take a look to the car's engine. He came, and made the attempt to start it, but nothing.

He looked around but found nothing wrong.

There were already 8:00 in the morning; I remembered the two previous times and I decided to repeat the procedure: wait. Time passed by and at about 8:15 I tried to start the car and as usual, the car started without a problem. This was beside the point, not to say that I arrived at my appointment, behind schedule. I explained, and at least; the commissioner understood my problem.

Someone said. "Third time's a charm." So when I returned to New York, on the first opportunity I had, and one more time, I went to the auto-dealer but they gave the same answer: NOTHING. They did not find anything.

"It was already getting dark." Frustration piled up. I was losing confidence in this car, and I didn't know what to do.

This experience seems that has nothing related to R & A/C but . . . hold it, it's *electrical* . . . And as we would see, it might happen in our field as well.

I experienced this problem for a long time. Approximately eight months.

Did I have a lemon in my hands?

Other questions

    1.   Do you have any idea of what the problem was?

    2.   What should you do?

Please, don't send me to Detroit. Sometimes, help arrives; from the least expected.

===

EXPERIENCE 126-A

Before reading this experience, we recommend trying to make a list of all the refrigeration mechanical components used in R and A/C systems, since any one of them may present refrigerant leaks.

This service call, according to the dispatcher, who said: "The client complained that the system was not working properly; because it had a lack of refrigerant." Up to this point, 3 mechanics had been sent, with no positive results. Each of the three times they reported, no leaks in the system. But . . .

I went to the place, met the customer who told me exactly the same history as the dispatcher. For me, if it is a lack of refrigerant it should be a leak of it.

I started getting familiar with the system. This equipment had the following components and characteristics:

1. It used a 3/4 of a ton semi hermetic unit, equipped with packing gland type, suction and discharge service valves.
2. It had a water cooled shell-and-tube type condenser.
3. It used R-134a.
4. Individual high- and low-pressure controls.

    a. High-pressure control was connected directly to the system's high side.
    b. Low-pressure control was connected to the unit's suction service valve.

5. A water-regulating valve.

    It was connected in the same place, along with the high-pressure control.

Note. Some systems may have or use, other mechanical components attached to it.

To install the suction gauge, the low-pressure control had to be disconnected, then, connected to one of the other manifold's hoses (*).

(*) Sometimes, while the work is performed some mechanics prefer to "jump-out" this control.

Based on the fact, that so many calls had been performed; with the same "leaked refrigerant" complaint.

Questions

1. What should we do, other than the other service people did?

2. Where can be located that leak, that the three mechanics; fail to find?

3. What specific test, should we perform to find the leaks?

==

EXPERIENCE 127-A

This service call was to a delicatessen; the client's complaint, the frozen food products; display cabinet "does not cool." The air-cooled condensing unit was located at the back of the store. Temperature there was 88°F.

Characteristics of the equipment.

1. Semi hermetic condensing unit of 3/4 of a ton.
2. Operating with 230 volts, 60 cycles and 1 phase.
3. Using R-502
4. Needed temperature 15°F.
5. It uses a dual (high and low) pressure control.
6. It uses a solenoid valve on the liquid line, controlled by a thermostat; within the cooled space.

An inspection to the equipment showed that the compressor has been working. At the time of the inspection, it was stalled. When installing the pressure gauges, the suction pressure was low (20 psig), which demonstrated that the unit was stopped by the system's low-pressure control.

===

Questions

1. What reasons may exist, so a machine with these characteristics; stops by a low-pressure control?

2. What another procedure is due to carry out, before expressing an opinion?

3. What should be the operating pressures, in this equipment?

EXPERIENCE 128-A

This service call was to a delicatessen. The refrigeration equipment had two rich-in show-cases; number one {1} low temperature (15°F.) for meats and frozen products.

According to the attendant, the number two {2} medium temperature (48°F.) used for milk products, vegetables, etc., it was not working well. And according to the customer, the products stored there, get frozen.

The refrigeration equipment had the following characteristics:

1. Semi hermetic unit 1 hp, 115 volts, single-phase, 60 Hz.
2. Two thermostatic expansion valves.
3. It used refrigerant 134a.
4. Condensation by air.
5. Both cooling spaces are provided with solenoid valves in their liquid lines.
6. The medium temperature evaporator, in its exit, has a pressure regulator (EPR)
7. A dual (high and low) pressure control.

Temperature at the condenser's equipment site was 80°F.

The service gauges were installed, and the operating pressures were verified; (High pressure = 125 psig Low pressure = 12 psig)

It was also verified:

1. The EPR operative condition. It was found adjusted at 10 psig (7°F.)
2. The two liquid line's solenoid valves; were energized and in normal operation.

3. The adjustment of the two pressure controls were:

   High-pressure control (Cut-out point = 175 psig. Cut-in point = Manual reset)
   Low-pressure control (Cut-out point = 0 psig (-16°F)
   Cut-in point = 8 psig (13°F)

4. The superheat, in both thermostatic expansion valves, was:

   a) Low-temperature application = 9.5°F.
   b) Medium temperature application = 5.3°F.

Questions

1. What should be our diagnosis for this problem?

2. What should be the operating pressures values here?

3. What should be the setting of the EPR?

===

EXPERIENCE 129-A

This former student called me, asking for some aid. One of his customers requires a work for an ice block machine.

He needed the basic information to calculate the heat load. (He required the information for only one ice-block container.) He gave me the following information:

1. Water temperature = 60°F (15.6°C)
2. Ice temperature = 18°F (-7.7°C)
3. Operating time = 8 hours
4. Ice blocks dimensions:

   a) 48" long x 48" wide x 18" thick

      4' x 4' x 1.5' =   24.0 cu ft
                      x 62.4 lb /cu ft (*)
                      = 1,498.00 lb of water

         62.4 lb/cu ft / 8.34 lb/gal (**) = 7.48 gal/cu ft (***)

(*) Weight (lb/cu ft) pounds of water per cubic foot.
(**) Weight (lb/gal) pounds of water per gallon.
(***) Number (gal/cu ft) gallons per cubic foot.

Questions

1. Do you have any questions for this requirement?

2. What type of refrigerant should we recommend for this application?

3. What should be the required operating pressures?

==

EXPERIENCE 130-A

By reference of a delicatessen owner, a former student was called to repair the commercial refrigeration equipment used to cool beer and other drinks.

These products were wholesaled among others to bars and stores.

According to the client, for the same problem, other two refrigeration service companies have been here. Both companies agreed in their diagnosis: To change the compressor! The customer said that the equipment was only four years old.

The main reason for this call was that for several weeks, the compressor of this equipment made a noise as if something was loose within. In addition, it did not cool properly.

The client says that he has the business for sale, and he doesn't want to spend much money.

This refrigeration equipment worked for two walking refrigerating spaces:

1. Cooling kegs of 25/30 pounds, to be used in bars, parties, etc. (32°F)
2. Cooling of beer bottles and tin cans (38°F)

This equipment had the following characteristics:

1. Semi-hermetic compressor "V" type. Capacity: 5 tons.
2. Operates with 440 volts, 3 phases, and 60 cycles.
3. Condensation by water provided by a cooling tower (82°F)
4. A receiver on the liquid line, with a packing gland type "king" valve.

5. Uses R-134a
6. Liquid, as well as moisture indicators on the liquid lines of both rooms.
7. Two solenoids valves on the liquid lines, of each cooled space. Thermostats within each space controlled these valves.
8. A dual-type pressure control.

Manifold was installed and gave the following readings:

a) High pressure 115 psig (lower than normal)
b) Low pressure 20 psig (higher than normal)
c) Running amperage (lower than normal)

Temperature readings:

a) Liquid line approximately 80°F (lower than normal)
b) Suction line approximately 50°F (higher than normal)

Questions

1. What do you think is the problem here?

2. What test or tests may you recommend to carry out to verify this unit?

3. Would you recommend to repair this compressor?

4. Which should be the high and low operating pressures and liquid and suction lines operating normal temperatures on this set?

5. What should be the temperature difference, among the condenser's water supply-return lines?

6. Besides the components already mentioned, what other mechanical device should be installed? And where?

==

Other related experiences

1. My Railroad Work's Experience-1
2. My Railroad Work's Experience-2
3. My HVAC & R Teaching Beginnings
4. More Than Just R & A/C Experiences
5. Other School Teaching Experiences

My Railroad Work's Experience - 1

It was in the middle of the year 1949 (I was fifteen years old), I just left Saint Antony (my youth technical School). I had to convince my father so that he helped me, to enter to work in the railroads; he alleged that I was "rebellious" about taking or following orders (remembering to me, the work in the laboratory of Mr. Cardona, when I was thirteen years old).

Finally, I could convince him; he moved all his influences with people he knew (among others, I believe, was with his cousin) and he was able, to get me accepted.

When admitted in the railroad, I was sixteen years old and I was not very athletic nor strong, that I can say. I was hoping to enter, and work in the shop where winches, lathes, drills, brushes and other sophisticated and bigger machinery, with which I had already worked with in Saint Antony. Or; in the locomotives, like I had always been my dream.

But they did not believe me (I think, my height did not help me). Instead, they sent me to work in the transport branch (rolled material). I guess, it should be the pattern of my life.

Wrong place, at the wrong time.

This work had to do with the repair-maintenance of passengers and load vehicles. I did not like it much, but my father told me: "Now, you have to accept it, because we already mopped ourselves enough, to obtain it." I did not have another alternative but to accept.

In 1950, I was assigned to work in the Northeast railroad shops; they were located in the Samper Mendoza borough; at the South of Bogota City.

My title was apprentice. I won (if my memory doesn't fail) about 67 cents per day.

I vividly remember my first day at work; I was assigned to work with a man, whom last name was Celis. We greet each other; then he said to me: "Put that jack on your shoulder." This, it was (for me) a heavy tool, of about two feet high; it was used to lift vehicles to repair them. It weighed I believe, about twenty-five pounds (for me, like one ton). I obediently try to lift it, with great difficulty and I didn't even move it. The man took it and placed it; on my right shoulder.

With great pain, I began to walk behind the man. In our way, everyone we found greeted him and he stopped to speak with whomever; until (thanks God), we finally arrived at our destiny. It seemed to me, like an eternity. When he issued me the order, to put the jack on the ground, I simply drop it. I could no longer, holder it anymore. That night when arriving at home, my shoulder hurt a lot and I had a sore on it; I did cure it as I could. No way to complain, since my father had advised to me that this work was heavy.

The following day was a lot better; I began to take the thread of the subject. On the following days, I learned the vocabulary and names of all the parts that composed the tools; and parts of the vehicles. I also began to carry out all types of jobs, without the boss having to tell me.

My regular schedule of work was from 8:00 to 12:00 p.m. and from 1:00 to 5:00 p.m. And at the end of the day, I always looked, for some overtime; with the idea, of avoiding problems at my home, I tried to be there, the shortest time as possible. To the effect, I found a job that it began at five o'clock in the morning, to wash the cars of the railroad's directors. As we lived in the North of Bogota (more or less to 1.5 hours of the work place, by bus), I rose at about three o'clock and I walked, five streets to the bus stop.

In order to make my transportation easier and faster, I bought a second-hand bicycle. I fixed by myself. It lasted several months until one night that I was traveling North-bound direction, thru 17th Avenue, a public service car (a taxi-cab) it went through the traffic red light, I didn't have time to stop my vehicle; and impacted against the car, thanks to God nothing serious happen to me or the car. The bicycle was destroyed. Battered and limping I carried it till my home (distant about 2.5 miles). In the following days, no matter how hard I tried to fix it, I could not do it; it was in a very bad shape.

I was working at the Samper Mendoza's railroad shops, until they needed a helper for the main Central station. Due to my dedication and good behavior, they recommended me. I was promoted to 3rd. class helper. Now the wage rose, I earned $1.33 pesos per day.

My new boss was a brown-colored man of name Isidro Peña, but they nicknamed him "—— Peña." According to what I heard, no helper last too long working with him; because he was, too demanding.

In my new schedule, I entered five o'clock in the morning, and finished at three o'clock in the afternoon, but I managed to get overtime and I arranged to arrive at home; more or less at about ten o'clock at night.

The work day at the Central station, began with the repair of the vehicles that had arrived the previous night.

The first day, at about ten o'clock in the morning; the boss taught me the "booklet."

Tools that I had to take with me were:

1. An oil can, with a long tip (a foot and half long) (capacity 2 1/2 gallons). I place it to the sun, so that the oil was lukewarm, and at the time of using it, it flowed easily.

2. An iron piece for "Stirring," with the flat end of about 3' of length x ¾" thick. This tool was used to fix the packing that was in each bearing box (rowlocks) of each vehicle's wheels.

3. Another piece of iron with a hook in its end, the same size of the previous one. This was used to extract the damaged bearing's packing.

4. Another important tool that it was used was a small iron bar of 3.5' length by 1" thick. It was used, to change/turn around the brake's shoes. I didn't have to take this with me.

When the passenger's trains arrived, it was necessary to initiate the revision-job by the end along and opposed side of the passenger's exit. The boss had in its hand a piece of chalk and was reviewing each wheel {each vehicle has two cars (trucks), each one with four wheels).

For each function to be performed, he had one mark, and he explained to me:

If oil was needed a "vertical line," for an adjustment of the packing an "oblique line," to change the packing a "circle," to change a brake's shoe an "X," for brake's shoe turning, an "X" with a horizontal line above it, and so on.

The first passenger's train it arrived, at one o'clock in the afternoon.

I did not lose nor a single step behind the boss. I finished this stage, almost at the same time than him. It was necessary to work fast; once the train began to be "disarmed," the work to be performed will turn but difficult; still I had to change or to turn around the brake's shoes and to finish other repairs.

For that part, I had to find the vehicle, and to make what were necessary.

Everything was well. Until, I had to change a brake's shoe; that was in a bad condition. I tried to change, by "all" means on my reach. But I could not do it. What I did "very wrong," was to erase the mark; leaving the damned brake's shoe without changing it.

I didn't count, that my boss was not an idiot; he was reviewing each mark that he had made and that everything had been carried out.

At about 2:30 in the afternoon, he came and he asked me: "Did you finish? I answered him "Yes." Then, he told me: "Take the small bar, that break's shoe and come with me."

I did what I was told, and I followed him. When arriving at front of the vehicle at issue, he told me to erase the mark, and said, "Do you know what can happen through this that you left off to do? This can cause a derailment and can kill people. Why didn't you do it? Or why didn't you warn me?"

The man was so furious that I feared for a moment that he was going to strike me.

I answered him fearfully, "Because I could not do it." Then he gave me the following lesson, for all my life:

"Human beings shouldn't say the words *I CAN'T*."

"Everything in life is possible; remember it!" Immediately afterward he said, "Change the break's shoe"; I tried and tried very and tried hard, I hurt my hands, and in the end, after almost 45 minutes, I was successful.

In regular days, we finished repairing the vehicles of the previous day, at about 8:00 or 8:30; then we were going away in a "Small vehicle" with a gasoline motor; to repair the vehicles that had been left in some place of the line, on the Northeast's railroad; far from Bogota, as much as 50 kilometers.

This gentleman never said to me, anything bad or good. By third people, I knew that he said sometimes: "This boy is the best assistant that I ever had in my life. He is going to get, very far in life."

===

My Railroad Work's Experience - 2

While in the railroad work, for a short time, luckily (I believe was due to Pena's vacation), I had to work with another mechanic; to whom all called "Slow cadence" by its parsimonious walk, and to make things. In order to change his clothes, he took about twenty minutes. As in the morning (five o'clock) it was very cold, he warmed up his sox, by putting them to its mouth and blowing in them.

As I was accustomed to work fast, Pena's way of working. For that reason, I always felt uncomfortable. But in life, things do not always go as they are desired. In an aim, it is necessary to comply as it comes.

—— Peña returned to work, and I returned to work with him. One day, a supervisor named Justo Silva approached me and told me that I had been recommended; and if I would be interested in a promotion to first class helper, that implied, a transfer to the city of Sogamoso in the Department of Boyacá (located at about three hundred kilometers of Bogota); that was at the end of the Northeast railroad; it was located near the iron and steel company named: Paz the Rio, whose facilities, were in its beginnings.

As I said before, my situation at home was not good; for that reason, I decided to accept the transfer. I had to put myself, under the prescribed medical examinations. One of those, it showed that I had; an obstruction in the central part of the nose (a tissue), and it had to be removed, by a chirurgical procedure. Without the removal of it, it was difficult for me to breathe. This procedure was performed at the Central Clinic, to the north of Bogota. Two days later, the doctor was sending me home (in convalescence) by two weeks. According to the doctor, it was a chance of a hemorrhage; when doing any effort. I did not accept, and I told him, that I better returned to work immediately. The doctor made me sign a letter releasing him of any responsibility.

I traveled to that city, and I undertook my work as of first-class helper, with a man named Luis Tobias Lopez; whom I had known it (as a policeman) several years before in another town (there, they know him by the nickname of ———). I asked him, but he denied it. I leave thus. In my new work, I earned a little more than five pesos per day. Wage normal in those days.

The work there was arduous; many times we worked continuously up to three days followed with its nights. As I said before, the iron and steel factory (Steelers) of Paz the Rio were in its beginnings; the construction equipment and materials arrived up, in three or four daily trains (of twenty and twenty-five vehicles each one) including Sundays; aside from this, we had to repair and to maintain a daily passenger's train of about 20 vehicles, the locomotives of each train as well as a daily express vehicle.

We worked in mechanics of all types (boilers, alternative machines, electricity, air brakes, smelting, bearing, etc.); we had to be recursive, the entire railroad's operations there, depended on us. The experience acquired there, was extraordinary. There, we even performed works, that required three or four days in the main shops in Bogota; we did them, only in hours. Besides, those jobs required special tools. Many times I applied the mechanical knowledge, acquired in Saint Antony.

In one occasion, a locomotive arrived with a loud noise on the rowlock, of the main right connecting rod. The locomotive machine operator refused to return to Bogota in that condition. From the main shops at Bogota, they said that we would have to wait for until them sent a new connecting rod, and that should take a couple of days.

One never knows that what you see or hear; can serve some time on life. While I was working at the railroad at the Northeast shops in Bogota, only to look around, I went to the melting shop, to observe the different processes; in which by using wood-forms of different pieces, they carry out the melting procedures of those pieces.

By using that experience, I spoke with Tobias and I suggested to him (explained) that we could fix it. We placed the connecting rod in the ground (on a "bed" of fine-dry sand), then, by using a flat tin piece, we made a "form" (simulating the space occupied, by the friction material on the rowlock); we collected (25 pounds) of "Babbitt," anti-friction material and using a crucible we melted it. We prepared, cleaned the rowlock, and by using melted lead, we gave a bath to the rowlock surface, in order to get the appropriate attachment to it, of the anti-friction metal. We spilled the antifriction metal in the space, between the rowlock and the tin piece.

The more difficult part (due to its size) was the action of "seat-to rectify" the rowlock in the wrist-union. This work took us about four hours to complete. The locomotive could return, without problems.

Another time, the machine-operator of one of the load-carry trains called, and advised us to be prepared for a repair on the locomotive; they were experiencing a lot of problems. The train that was expected about two or three o'clock in the afternoon was behind schedule, it arrived; approximately about 7.30 at night. The problem consisted of, by a negligence of the fireman in the provision of water to the boiler, one fuse-plug —— within the boiler's fire-chamber had been melted, and the steam was leaking into the combustion chamber, interfering so, with the locomotive's normal operation (force-power). The piece had to be removed it, repair it and returned it to its location. The steam engine was shut off a few minutes after the arrival. The temperature within the boiler space was (4 or 5 atmospheres) between 454 and 650°C (850 and 1,200°F).

Only four hours after the locomotive's fire was extinguished, they made me enter the boiler's combustion chamber (it was do it through the opening of the boiler, that had a space but of 15" of diameter) it was hard for me to enter through it. The fuse-plug was located in the ceiling at the center of the fire-chamber and to remove it was necessary to use a special wrench; to complete the job I went in and out through that opening, no less than twenty times; the heat was suffocating, something outside this world; they wet rags and they hurled to the center of the combustion chamber, with the intent of lowering the temperature.

The rags were dried almost right away they throw them in. The threads of the fuse-plug were very fine (fact that no one knew) and because of that, it was necessary to turn it, many times; I believe no less than thirty times. In the end and suddenly, when the fuse-plug came out; there was a tremendous steam spurt and I had only time to through my arms through the opening. They had to, making a great effort, and after a very anguish moments, pull me out from the firing chamber.

I passed out. Half an hour later, I recover and I found myself all covered in many blankets and I was sweating profusely. It did not felt the legs, but until much later (near 2 hours; by then, I had my legs bleeding; with the worry, I was tweaking-pinching them). Here, as always, I thank God that I did not have later problems.

A few days later, after the job completion, we learned that the main shops in Bogota wait a couple of days before them intent to perform this type of job.

===

My HVAC & R teaching beginnings

On board of the frigate *Admiral Padilla*, the refrigeration system continued working well; I managed to use my recently acquired knowledge (through the books), to keep it operational. But I was not very convinced of which I was doing.

Returning from a cruise, I learned that the navy's technical schools in Barranquilla were offering the first refrigeration course at 560 hours (approximately five months).

I didn't have any wish to go, because of the instructors; they did not convinced me (Chiefs Moreno and Gomez) as well as, I had the idea that the first course was not always the best one.

Nevertheless, the ship's commander ("considering my achievements") recommended me, so I did not have other alternative but to go.

At the beginning of 1960, I travel to Barranquilla to take the refrigeration course. The course director, was a ship Engineer Lieutenant Hector Baron Rosas and instructors, sergeants Majors Moreno and Gomez. In this course, there were 14 students, petty officers with ranks of first, second and third class. My rank then was seaman. I was the only one with that rank in the course.

Wrong place, wrong time? I didn't think so. I enter the navy, when I was almost twenty-two years old.

Lieutenant Baron dictated some classes, but most of them were dictated by both Sergeants Moreno and Gomez.

The test book used, was Chapter 59 of the Manual of the Bureau of the U.S. Navy, in a version translated to the Spanish language; in other Spanish speaking country. In the translation, jargon typical of the country was used, —— was said to refer to the switches and other names of the sort; others parts of the translation were "miserable." I was struggling to get the best benefit of it but, I decided to make my "own translating version."

In the base's library I was able to find the only chapter 59 in the original English language version, I borrowed. I had some experience with this book, which I acquired on board of the Admiral Padilla. (I learn from the librarian, that the chiefs had removed from the library, all the English books. They missed the one I found).

Already accustomed to translating, I got a dictionary, and I worked with it.

Sergeant Moreno never used the board (not even to write his name); its classes consisted of saying, "Open the book in page _____" then he said. "Jimenez, read." I did it. When finishing, he asked, "Does somebody have a doubt or a question?" It was almost always a student who had a question or was not sure, and let the sergeant to know; then, he commanded me to read again. If still, after the second reading, there was a question; then, he commanded the student to make "20 push-ups."

He alleged that the student was "sabotaging the class."

Thus, we were in this tedious task for some time, until a day; I already had prepared myself and it asked the chief, if I could go to the blackboard, and explain the operation of the water regulating valve (it was the class that it corresponded to that hour) he said to me yes. I went to the board and "taught" the class.

Sergeant Gomez had a better will, but the classes with him did not arrive anywhere either. To me, he looked "lost." At one time, he asked me if I wanted to go with him to do a job (outside the base). I said, "Of course!" It seemed to me a good opportunity for "gaining" some practical experience, and I went with him. The first thing that he did when we arrived at the place was "calibrate" the low-pressure gauge for the refrigerant type in the system.

He asked me, what refrigerant it said in the system's ID plate. I look at the plate, and answered to him "R-12." He removed the low-pressure gauge's glass; then, by using a screwdriver, he proceeded to adjust the needle to the "0" temperature point reading of R-12! I asked him, "Why do you do that?" And he answered me that it had to be done whenever it was going to work with each particular refrigerant. I put no argument. It did not want, to expose myself to any misfortune.

Every week, the course had an examination. Usually, Sergeant Moreno prepared the test. Big problem. On some of those tests, more than 50% of the test questions, the chief Moreno pretended that we "memorized" the Pressure/Temperature Table (this, is a table with hundreds of pressure/temperature relationship values, for several refrigerants).

I continued dictating the classes of sergeant's Moreno. By that way, it continued the course, until its conclusion.

At the end of the course, I got the first place. Then, Lieutenant Baron, recommended to the base commander, to leave me, as an instructor's helper.

The refrigeration course number 2, began. Among the students, there were 2 sergeant majors, as well as petty officers of first, second and third Class. Completing the number of 14. As it would be continued with the same instructors, the same test-book that was used in Course number one; was used in this one.

Things continued the same, sergeant Gómez, dictated some classes. The "problem" was that he discovered the Chief Moreno's game of relaying to me the teaching, and he decided to do it as well. I did not put any objection; I considered it for my benefit.

This way, it continued the course; every day the sergeants had something to do and I was called to dictate the classes, I prepared, presented/displayed them, to the best of my ability and possible way.

One month after the beginning of this course, Lieutenant Baron's wife, who was pregnant, was ready to give birth. When this event happens; the husband takes a month off duty.

Before leaving, Lieutenant Baron called us to the office and distributed the classes. He assigned to each of us—Chief Moreno, Chief Gomez, and me—the classes that we had to teach (during his absence). My part of the teaching was "Tools."

This beside the point to say, that the classes continued like until now. The sergeants, rarely appeared in the classroom; they had always something else, to do.

The day Lieutenant Baron returned, the students requested "chain of command"; to speak with him. They communicated him, what it had happened in its absence (that I had been the only instructor, all the time) then, they solicited to him that it let me to be the sole instructor, for the course's remain.

Both sergeant Majors were sent to the Naval Base of Leticia (in the Amazons) and I stay as sole instructor. I never intended to cause any problem to neither of them, on the contrary; to me, they made me a favor. In addition, they knew what they we're doing. They did not count out, that; people cannot be deceived; indefinitely.

===

"More than just R and A/C" Experiences

The year 1958 was called the "Geophysical International Year."

While in the navy for this other trip, a scientist commission of the University of La Holla (San Diego, California) arrived. We left Cartagena and sail; we passed the Panama Canal; once on the Pacific Ocean, we took North direction, and we made coastal navigation (parallel to the coast). This it was a very interesting trip.

Throughout my life, I had been exposed to different situations. Through them, I had acquired some experience, in different fields, which at some time I had used them. Here, I would like to comment about one of them.

After a couple of hours of sailing in North direction, and at a specific time, the scientists performed a varied set of experiments; the maneuver went as follows (See arrows and graph below):

One hour in straight line, during which (The Latitude & Longitude coordinates were taken), then, a 90 degrees turn to the left was performed; sailed thus by 15 minutes; passed this time, make another 90 degrees turn to the right, sailed by 15 minutes; passed them, another turn of 90 degrees to the right, sailed thus by other 15 minutes. At the end of these 45 minutes, another turn of 90 degrees toward the left; returning, to recover the initial route.

During the 45 minutes maneuver; the scientists by using pertinent instruments, carried out the following experiments:

1. Water flow direction.
2. Using the "Sonar" they investigated location's *depth*.
3. Using a bucket; a seawater sample was taken, from it, they had its *temperature, salinity, density* and *specific gravity*.

Same procedures were performed with the air:

1. Direction.
2. Speed *(anemometer)*.
3. Temperatures *(dry bulb, wet bulb, dew point)*
4. *Relative humidity, specific volume, enthalpy or heat content (Btu), etc.*

These experiments were carried out during several days and at different hours. In the end, we returned to Cartagena.

When we were sailing, I liked to serve the 4:00 a.m. to 8:00 a.m. shifts. At that hour, (also in the 12:00 p.m. to 4:00 a.m. shift) the duty Officer in the bridge, allowed us the use the sextant (navigation device), used for "lowering stars" and to draw on a map the ship location; in addition, it was very interesting to see the sunrise.

It was interesting as well, "to compare" both the Atlantic and the Pacific oceans; the first one rarely serene, always has waves and the ship is in constant movement; the second makes an honor to his name, most of the time is like a mirror; at night, the stars can be seen reflected on; also, the dolphins can be observed, side to side of the ship, so seem to accompany the ship all along. In addition, in the Pacific sea, the tide can be observed; in the afternoons it can allow a person to walk out into the sea (beach), as far as a distance of two a three blocks, depending on where one is located.

===

Other School Teaching Experiences

One more time. Here, I don't try to judge anyone. I only narrate, what it was my own experience.

We are in the month of May, having my Technical Trade school closed; one morning (about 10:30), rather by boredom, and to occupy myself in something; until the new school's owner, where I was going to work, decided to called me. I took my credentials and I went to Brooklyn to another trade school.

Here, all the classes were in the English language. I went to the school's director office and I got an interview. I showed him, all my credentials. He was impressed with my resume and he told me that I would begin the following day. While he was showing me the school, the owner passed by and grudgingly said; "I do not pay more than $10.00 per hour."

I did not put any attention to what he said; I always had the idea that our work, in any field, it is a sale product and we must show what we intent to be worth.

At about 1:30, he introduced me to an instructor (Italian) that was teaching a class of the National Electric Code (NEC). I sat and stayed in its class, to observe: It was a class of about 30 students. Some of them were sleeping, others were reading the newspaper, or speaking among them; others left and enter the class like, they were in their own house; as if the instructor did not exist or were there.

While the class was in progress, I observe that the instructor had an Electrical Code book, with many colored marks on it. I guessed the most relevant points. Later on, the director gave me the same, but brand new

NEC book, and told me as well, this will be the book that I would use in my classes.

At about 3:00 p.m., at break time I determined that the observed class was enough for me. Since I didn't have any teaching experience in this subject, I approached the instructor, and I asked him if he could lend me his book; so, I could copy or mark some of the important sentences in it.

He looked at me and he said: "Jose, This book is like my life; but, you inspire confidence to me, and I am going to render it to you. Take a very good care of it" and, he gave me the book. That afternoon, before arriving at home, I went to a bookstore and bought a set of colors markers. That night, I was until high hours of the night trying "to copy" all the marks in my new book.

The following morning my first teaching class in this school. I arrived at 8:30 in the morning. I'm accustomed to always arrive on time to all my appointments, thus to be prepared.

I wrote my name on the blackboard.

The students began to arrive on time; according to the attendance list they were about 30 of them. At nine o'clock, I called the attendance; and when I was ready to begin the class (as I observed the previous day), some of the students, prepare themselves to sleep, to read the newspaper, exchange ideas, in general, everything that I saw they did the previous day, as well, as some of them leaving the class.

As in those days I used canes as walking aids, I took one of them, and I gave three blows on top of the writing desk. The students jumped, very surprised; they watched me, and I begin to speak. I said to them! Gentlemen! I do not know the experience that you have had in this school.

I'm here and I'm not a decorative figure. If somebody wants to make another thing than to attend the class, then, let me know it, now. If anyone leaves or planned to do it, I do not expect he returns.

(While I spoke, one of the students, requested permission to go to the board, I granted it).

I continued and I said to them: "None of you know me, but while I am your instructor, I deserve respect, and beforehand, I want to know what is, what you wanted to do." By some reason, I have been assigned to teach this class.

I spent quite some time (I believe about twenty minutes); the student at the board had finished writing. He wrote about thirty names. Then he said to me "Mr. Jimenez," I have been in this school for the last six months, and these are the names of some of the instructors who I have had, and the truth, I am not satisfied with the instruction."

At this moment, I find out that in this class, there were students with different time frames in the school: six, four, two months, etc., including some students who began in this day.

I continued: "If you think that I do not have the knowledge, or what you need, you can go to the school's director and express your problem. Or you can wait in the class to continue and when I finish it, then you can make your decision.

Needless to say, that they all took their seats, and they were prepared to listen to the class.

It had passed two weeks, everybody has something to say of my classes; the students, don't want to lose nor a minute of them. I even had the visit of other students from other classes that requested permission to attend mine. I receipt my first check and, sure I knew that was going to be of ten dollars per hour. As everybody in the school, already know me, I decide to speak with the school's owner. I went to his office, and with a smile of "ear to ear" he answered my salute, and he asked me; what it must be my visit.

I said, "Sir, the reason is to ask if you are satisfied with my work." He answered me, "Oh yes, Mr. Jimenez, very much." Then I said to him that I was not, with my payment. That I need an increase, because I believe it that I was worth, much more. With a funny smile, he says to me: "Oh, I know Mr. Jimenez, and I am going to take care of that."

In the following days, my success continued. I made a material requisition, tools/equipment (and, by the surprise of all) I was supplied immediately with what I had solicited. The owner called me to his office and he asked me to teach (at nights from 7 to 10), in the school that he had in Manhattan. I accepted.

Another two weeks had passed and I received my check, with an increase of $1.00 per hour. Only to let him know what I thought, I went to the owner's office, and knowing what I wanted, he asked me, "What's the matter now, Mr. Jimenez?" I answered him, "Money!" And he, with a funny smile, said, "I know, it's OK, Mr. Jimenez. I'll take care of that." I left his office but not convinced at all of his promises.

The school in Manhattan had 5 floors (no elevator); I told them, if they want me to work there, I will do it but in the first floor. They agreed and assigned a place, equipment and materials. I continue working in both places and everybody, but me, was pleased.

The second paycheck arrived; this time with an increase of $0.50. As I thought, that with the owner I was going to get nowhere, I decided to go to the director's office and to speak with him. I told him the problem, and I warned him, that finishing that week I would quit. He asked me to let him speak with the owner. I advised him, that he was not going to obtain anything. That will be a waste of time.

My quitting news, spreader very fast, and the students came to me to request, that please do not leave now; that they were learning and were so happy. Only because of them, I promised to continue, until finishing that phase of the training course. I let the director know so. I also told him that I continued but only in the Brooklyn's school. The three weeks passed, and I left.

===

EXPERIENCE 01-B

<u>Procedure</u>

I requested that they charge the system, with enough refrigerant (R-12) and put it to work. The operating discharge pressure was approximately 125 psig, good enough for high side-leak detection. With the equipment in operation, I opened a valve in the condenser's water outlet side and collected some of the condenser outlet water; in a small bucket. After passing the halide torch hose over the surface of the water, my suspicion was correct. The change of the leak detector's flame "blue" color when normal, to the "green" color indicating a leak, showed refrigerant was present in each of the two vessel's water samples. The problem was located in the condenser's tubes.

Both captains asked, "What can you do?" and "What do you need?"

They were under extreme pressure and told me they would supply me with whatever I needed.

I told them I would need two crewmembers to help me, and they offered me their entire crew of 25 men in each ship.

I made my tool list and handed it to the machine room chief:

> One nitrogen cylinder with its pressure regulator.
> Refrigerant cylinders to store the R-12
> One vacuum pump
> Gauge manifold with hoses, wrenches, etc.

===

Now, my plan:

First, completely drain all the water out of both condensers.

Next, pump-down the system to remove/recover all refrigerant from each system into empty cylinders until a reading of only five psig were left in the system. This process was limited to one system at a time, because they did not have enough cylinders for the two systems.

After removing the refrigerant, we proceeded to isolate the condenser from the rest of the system, by front seating the compressor's discharge packing gland service valve and the condenser outlet liquid line service valve (aka king valve).

Now, we carefully made a mark on the condenser head ends to ensure we would return them to the same position. Using differentials, we removed the condenser's heads with extreme care so as not to cause damage to the gaskets.

Using nitrogen, we blew through each tube from one end to remove any water vestige from it and avoid the entrance of moisture to the system.

If my memory serves me correctly, they were about 250 tubes of approximately 5/8" in diameter. The maximum number of tubes that we could seal and still operate the system would be 10% or with this machine, 25 tubes in each condenser.

We needed to plug the 250 tubes. We tested several sizes of cork. We determined we would need a cork within the tube, and outside the tube with an approximate length of 1 1/4" to 1 1/2." This would allow them to be removed without damaging them. Thus, when the appropriate size was found; 500 corks of that size were ordered. 1000 corks in total were obtained, relatively quickly.

As soon as the corks arrived we began plugging each tube on both sides, using a rubber or wood mallets until all of the tubes were sealed at both ends of the condenser. This was not an easy task.

After several hours, we were ready to test the tubes. We began by using the pressure of 5 psig of refrigerant R-12 left in the system. Then, using nitrogen through the regulator we carefully raised the pressure (step by step) to a maximum of 150 psig. Everyone understood that at this pressure, some of the corks would pop out, and we were hoping it would be within tolerance.

Next I suggested cutting some tubes, pieces of 5/8" diameter and 1.5" in length were pinched and sealed at one end, ready to be welded into any tube that displayed a leak.

Luck was with us when the first condenser was pressurized, and only nine corks popped out. We were well within the calculated tolerance.

We vented the pressure, completely bringing the system to atmospheric pressure. We silver soldered the pre-cut pieces of tube with oxyacetylene torch and sealed the nine leaking tubes at both ends taking care not to burn the corks still in place. This process took some time, allowing the welded parts to cool to room temperature.

Now we repeated pressurizing the system to 5 psig with refrigerant R-12, and then using nitrogen through the regulator we slowly raising the pressure to 150 psig. Refrigerant R-12 being heavier than air, we began using the leak detector at the upper part of one side of the condenser, very slowly, removing corks one by one to identify any leaks and if found, mark them. Two more leaking tubes were discovered. These tubes were sealed. The final leak check included soap bubble testing of the sealed tubes proving the system was tight and leak free.

The condenser of the first system now was evacuated using 2 vacuum pumps to remove the air, the moisture, as well as the non-condensable gases. The two head-ends were placed in their places, according to the marks done when they were removed, and after approximately 2 hours of vacuuming the system, we charged the refrigerant and put the system in normal operation.

As mentioned before, the exact same procedure was performed with the condenser in the second vessel. 11 tubes with leaks were discovered with the corks, and 2 more tubes were found using the leak detector, for a total of 13. This too was well below the maximum of 25 tubes, and so we could repair it.

All the work took a little more than eight 8 hours. The entire operations of both systems were verified and both captains said, "Mission accomplished."

At the end, they did not know how to thank me, so they filled a plastic bag with cigarette packages and went away very thankful. I was satisfied with the work we did.

Something that impressed me most was that for a moment I saw that the entire crews of both ships made me feel like I was in charge and did everything I suggested. Well, I never issued any order; they were, as I say, "suggestions."

===

EXPERIENCE 02-B

<u>Analysis</u>

According to the information compiled from the person in charge, and our own findings; the original call "Not enough cooling," was due to a dirty condenser.

Probably the first mechanic removed some refrigerant.

The second mechanic added some refrigerant.

And the third mechanic added refrigerant and adjusted the low-pressure control.

<u>Calculations</u>: *(R-12)*

Condensing temperature = Ambient temperature   = 80°F.
+ Temp. diff. (Δt - Dt) = <u>25–35°F.</u>
105–115°F.
Discharge pressure = 125 to 170 psig.

(B. P.) Boiling temperature = Required temperature = 40°F.
- Temp. diff. (Δt - Dt) = <u>15°F.</u>
25°F.
Suction pressure = 24.6 psig.

Low Pressure Control setting = Cut-in 35 psig; Diff. = 20 psig = Cut-out = 15 psig.
(*) Cut-out is ⟶ Cut-in minus ⟶ Differential.

<u>Procedure</u>:

I installed the pressure and found the high side pressure to be too high 190 psig. What the unit had was an excess of refrigerant.

By verifying the system's operation, also found the LP control's settings

Cut-out point = Settled too low. Differential should be 20.0 psig.

Solution

The condenser was cleaned. The excess of refrigerant was removed. The Low pressure control verified-adjusted and the unit returned to work without problem.

A few days later, the cafeteria's person in charge called me, this time to meet this man who was the sales representative of the beverages factory. He thanked me for the repair, offered me the job of maintaining and repairing all the beverage coolers (about 80) that they had in different locations; such as Stadiums, Schools, Colleges, etc., throughout the city.

===

EXPERIENCE 03-B

<u>Analysis</u>

By using the "X1" Ohm (Ω -1) scale in an analog instrument, the resistance/ ground conditions of the entire equipment was verified.

01. *The ohmmeter leads were connected to the "L1" and "L2" terminals at the selectors switch inlet.*

   Note. It was made sure, the selector switch was in the "OFF" position and the thermostat switch set on the "ON" position (closed).

02. *The ohmmeter gave an "∞" infinite (no resistance) reading.*

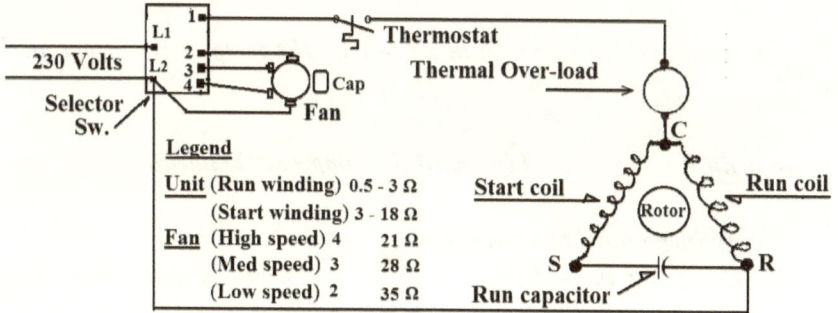

03. *The selector switch was moved to fan "Low" speed position, the meter read = _35 Ω_*

04. *The selector switch was moved to fan "Med" speed position, the meter read = _28 Ω_*

05. *The selector switch was moved to fan "High" speed position, the meter read = 21 Ω_*

06. *The selector switch was moved to "Cool" fan "Low" speed position, the meter read = _1.70 Ω_*

07. *The selector switch was moved to "Cool" fan "Med" speed position, the meter read = _1.69 Ω_*

08. *The selector switch was moved to "Cool" fan "High" speed position, the meter read = _1.66 Ω_*

As shown, the obtained resistance readings were normal and no ground reading was registered. All the other electrical components (selector sw., capacitors, thermostat, etc.) were verified as well and everything was in order.

Procedure

Since all the electrical components in the equipment were in good working conditions, the system was energized.

The fan motor worked normally in all speeds. The compressor did not work at all.

By using different methods, I intended the compressor to start. Among them:

1. *Adding a starting capacitor.*
2. *Energizing the unit in such way that it would reverse polarity and free the pistons.*
3. *The compressor was energized applying the power directly to it. But nothing worked.*

The compressor was stuck, probably broken (*), and it had to be changed.

(*) I got this suspicion, based on past experiences, with these types of compressors.

Then he put me in charge of repairing it, with a replacement that he would send from the United States.

Solution

I kept the equipment in my shop and two weeks later, the replacement part arrived. It was changed and once verified, was given to the new owner. Case solved.

Some time later, the lieutenant communicates with me asking for the A/C's repair cost. I did not charge anything; I owed him for helping me with the trip to take courses in the United States and having me in account for the repair of the coast guard ships (Experience 01).

==

EXPERIENCE 04-B

<u>Analysis</u>

I disconnected and submitted the unit to all types of electrical tests:

I assumed the electrical connections as: (See diagram below)

* 1. Start terminal.   * 2. Common terminal.   * 3. Run terminal.

And although everything (all the electrical components, etc.) showed well. By using different methods: a starting capacitor, a reversing procedure, high voltage, etc., I tried to make it start, but it did not work. Something was wrong with the mechanical part of the device. The compressor was stuck and it had to be changed.

I went to the beverage factory manager's office and I consulted the problem with him, then he said to me that I had removed that compressor long before I was going to use it and that I had to take care of the problem as best as I could.

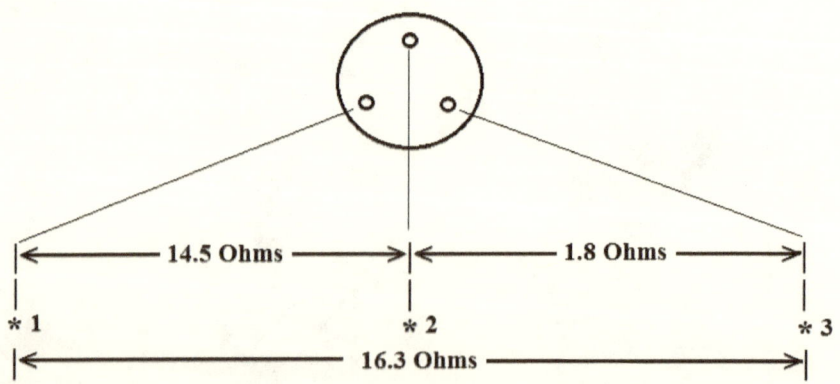

I asked the company's warehouse clerk, where they had bought those compressors and he gave me the name/bill number and the address of the store.

With the defective compressor, I went to the place and the employee to whom I talked said that the guarantee had expired. I took into account that with this man, I would not go anywhere. So, I requested to speak with a supervisor and explained to him the problem; with this man I did not have much problem. He ordered the clerk to give me a new compressor, and that was it.

I finished the installation and the new compressor worked well. Nevertheless, I got the idea that they knew of the problem with this compressor. I found the answer several years later. Read the following detailed information.

==

The following information was/is true to the best of my knowledge.

Manufacturers, in general, receive orders to fabricate a certain number of components or parts. For example, a compressor manufacturer receives an order to build all together 10,000 units of different capacities (2,000 of each denomination: 1/8, 1/4, 1/3 and 1/2 Horse Power). He normally produces 12,000 compressors. The surplus (excess) has the purpose of replacing the compressors for whatever reason or faults, in the 10,000 units ordered.

The 2,000 compressors of excess are stored and when a unit fails, or it is returned (for any reason: Mechanical-electrical-Etc.) they replace it with the ones made in excess. The returned compressor is stored and separated from the initial harvest. Usually, the Manufacturers guarantee lasts 2 years. At certain time after this expiration date. The companies send personnel to sell these units.

To us, my partner and I, a manufacturer's agent offered 5,000 units at the cost of $10.00 each one. Of course, we did not have this amount of money, but companies as "No Name" buy them, store them and then "sell" them at retail or in lots, as is; at the present day price, in the market (Example: Then, a compressor of 1/8 hp was worth $80.00). No Name had branches in other states of the union. I.e. in Miami, they came from South America and the

Caribbean and they bought those products with a discount of a 25% off the regular price.

Many of those units were in good operative condition, but some had defects and when installing, they showed up. I personally had this experience (here and abroad) with several units that I bought in and got from these warehouses.

The unit in my experience was one of those. I learned, that when the compressor I had a problem with was returned to the store, they opened and found the screws that secure the motor's stator to the motor's frame, were loose, in other words they were never tightened. That was the reason for the compressor NOT starting.

The service technicians should learn to read (interpret) the information given in those (ID) plates on each product. This way, we can know (among other things), the product manufacture's date.

==

EXPERIENCE 05-B

<u>Calculations</u>: *(R-22) Normal readings.*

According to the system's location and requirements; this equipment should have the following operating characteristics:

Condensing temperature = Water temperature   =   80°F.
*   + Temp. diff. (Δt - Dt) = <u>25–35</u>°F.*
*   105–115°F.*

Discharge pressure per the PT chart = 210–243 psig

Water temperature (In - Out Condenser) 10/12° = 90–92°F
*   Actual difference = 6°F*

(B. P.) Boiling temperature = Required temperature = 72°F.
*   - Temp. diff. (Δt - Dt) = <u>35–30</u>°F.*
*   37–42°F.*

Suction pressure = 64–72 psig.

Suction line temperature = Approx. 55–65°F.

Ambient temperature = 80°F.

Liquid line temperature = Approx. 90°F.

Running amperage was low, because of the operating conditions.

<u>Analysis</u>

As mentioned in Part "A," the system's operation was observed, through the manifold gauges, different tools and methods:

1. *Pressure readings.*

    a. *Discharge pressure = Actual reading 190 psig. Low.*

    b. *Suction pressure = Actual reading 48 psig (24°F). Abnormally Low.*

2. *Temperature tact feeling (by hand).*

    c. *Suction line temperature = approximately 20°F. Too low.*

    d. *Liquid line temperature = approximately 85°F. Low.*

The conclusion-suspicion was reached: Erratic work of the thermostatic expansion valve.

Carry out the suitable procedure (Pump-down).

The TX valve components were studied in great detail; it was observed that the three (3) pins, located below the valve's diaphragm, which work consistently by pushing the valve to open, "were flattened" (shortened) by the prolonged use (it was calculated that the shortage was about 3/32"). It was necessary to change them (since the parts were out of production).

With no alternatives to obtain then, it had to figure out, other way to fix the problem.

Solution

The TX valve's diaphragm surface, where the pins seated, were measured. The diameter was of approximately 1." In those days, in Colombia silver currencies of 50 cents were used; they were more or less, the diameter size of the diaphragm's surface; and a thickness of about 3/16." By using sandpaper, the thickness of the currency was reduced until what I the "thought" was the required size of (3/32") was obtained.

The valve was assembled using the coin in the appropriate place; checked the system for refrigerant leaks, pulled a vacuum procedure for about 1 hour and put the equipment back to work, charged with refrigerant and verified its operation. It worked without any problems.

Time of the repair, about two hours.

===

EXPERIENCE 06-B

<u>Analysis</u>:

Some refrigerant was added and the inspection showed a refrigerant leak in the lower part of the storage box, at the point where the refrigeration lines, (Suction and Liquid), come in-out of the evaporator attached to the boxes inside wall.

It was assumed that after all the refrigerant leaked out, as the system was left operating, it created a vacuum; and this effect absorbed the water collected in the leaking point and stored it in the compressor.

<u>Calculations</u>: *(R-12)*

Condensing temperature = Ambient temperature = 85°F.
+ Temp. diff. (Δt - Dt) = <u>25–35°F.</u>
110–120°F.

Discharge pressure = 136 to 157 psig.

(B. P.) Boiling temperature = Required temperature = 10°F.
- Temp. diff. (Δt - Dt) = <u>15°F.</u>
-5°F.

Suction pressure = 6.7 psig

Procedure

The compressor was disconnected from the system.

Solution

Once the compressor was disconnected, it was turned upside down to remove the oil; with this action, no less than 3/4 of a gallon of water came out of it.

By using a proper method, the oil in the compressor was changed three times; until finally the "grounded" condition disappeared.

The filter-drier on the liquid line was removed, and replaced with connectors and flare nuts, by using nitrogen, the piping lines were blown/cleaned. This showed an incredible amount of water coming out.

Then, the compressor was connected to the system again; after checking for leaks we found none. While the system was vacuumed for about 2 hours, light bulbs were placed inside the box to help to the evaporation of the water remaining in the evaporator.

At the end of the described procedure, and by using the vacuum in the compressor, oil was added to it. Then, refrigerant was added until the pressure rose up to 5 psig. At this stage, while refrigerant was leaking, a new filter-drier was connected in its place. The connection was verified for leaks and the system was charged with refrigerant.

==

EXPERIENCE 07-B

<u>Procedure</u>

I recommended he bring the compressor to my shop. He did it. After being disassembled, using the appropriate tools, it was checked out:

   a.   *The cylinders were OK.*
   b.   *The shaft was in good condition as well.*
   c.   *The connecting rods were in good shape but if possible I recommended getting a new set (six) anti-friction casts of babbitt (metal, of the removable type). One cast piece was removed so the captain could give the supplier a better idea.*

       The conclusion was reached, that the compressor needed both types of piston rings (Lubrication and Compression), in addition it would be advisable to change the valve's plate assemblies; as well as the complete set of gaskets.

   d.   *A complete set of one valve plate assembly was given to the captain as well.*

The necessary dimensions and information were taken, and given to the captain. He said he was going to try to obtain those spare parts, in Aruba or in Puerto Rico.

Two weeks later, he called to let me know that he had obtained the other parts, but not the rings. In the meantime, I was making inquiries and somebody told to me that it was possible to try the rings used in the automobile industry.

The problem was the ring's precise size; also, it was necessary to take into consideration the hardness of such rings. I consulted with an industrial mechanic, a friend of mine named Luis, (my wife's cousin), and he said, he thought that it was worth the trouble to run the risk. He told me, that he could help out me by obtaining them.

Solution

The captain bought the spare parts. We changed every part, by using the best of our knowledge. We performed the proper adjustments, (Connecting rod's friction castings), we had some problems in adjusting the compression-lubrication rings, but finally, the repair was completed.

To finish it and to verify the system's work, I went to the ship, inspected the installation and ran a test. Everything was perfect. Between the initial inspection, getting the spare parts and the final test and back to operation, it took us about 4 weeks.

===

EXPERIENCE 08-B

Procedure.

After rendering my report to the customer, we reached an agreement on the price. I had to stay a pair of days in a hotel until we finished the repair.

This man had all the spare parts and practically everything was at hand. There was no problem in performing this work.

All the electrical as well as mechanical components of the system were reviewed. Everything looked in proper order.

Calculations *(R-12)*

Condensing temperature = Seawater temperature =   80°F.
                    + Temp. diff. (Δt - Dt) = 15–25°F.
                                              95–105°F.
Discharge pressure = 108–126 psig.

Water temperature (In - Out Condenser) 8/10° = 88–90°F.

(B. P.) Boiling temperature = Required temperature = - 10°F.
                    - Temp. diff. (Δt - Dt) = - 15°F.
                                              -25°F.
Suction pressure = Approx. 0 psig.

Conclusion

After completing the repair, the system was assembled, checked for refrigerant leaks, all controls were adjusted, a piece of 5/8" tubing was placed in replacement of the liquid line filter-drier. The entire system was verified for leaks; the vacuum in the compressor was used to add the oil charge to it. Refrigerant was added until the gauges showed 5 psig, and while refrigerant was leaking through the filter dryer connection, the new device was connected; afterward, a check for leaks was carried out; adding the refrigerant charge completed the job.

===

EXPERIENCE 09-B

<u>Procedure</u> (Pilot valve's operation)

Normally, when the pilot valve closes and no more pressure is applied to the piston; in order for it to return to the closed position; the pressure on top of the piston has to be released; to accomplish this, the piston has a very small orifice. The refrigerant in the piston's cylinder passes through it and the piston, pushed by a spring, goes back to the closed initial position.

Solution

As I knew the operation of that valve, I asked the mechanic "did you clean the orifice on the piston's head?" He did not know what I was talking about; I explained to him that, in order for the piston to return to its "closing" position; the orifice allowed the equalization of pressures (Liquid line/ Evaporator) necessary, so the piston could return and cut the refrigerant flow. Otherwise the valve will remain open, producing the refrigerant flood-back problem. Then I left.

A few days later, I met with the mechanic and I asked him what had happened there. He answered me that it was what I had said and that the machine was working well.

===

EXPERIENCE 10-B

When arriving at the place, the compressor although working, the evaporator does not "show" frost, neither the peculiar sound (hissing-gurgling) indicative of the passage of refrigerant through the capillary tube boiling in the evaporator.

Diagnosis

According to the symptoms displayed here and hard to believe, I suspected the problem was "moisture in the system." After five years of normal operation?

Procedure

A plate was removed to allow the access to the place where the capillary tube, entered the evaporator.

By using a match, heat was applied in this point; immediately it felt-hear like the ice deposited at the outlet of the tube melted and the sound was heard (hissing-gurgling) of the refrigerant passing thru and therefore boiling and the cooling effect appears on the evaporator's surface.

Conclusion*: The drying agent (desiccant) lost its effect of "retaining" the moisture and it loosen letting it circulates with the refrigerant through the system.*

<u>Calculations</u> *(R-12)*

(B. P.) Boiling temperature = Required temperature = 32°F.
$\qquad\qquad\qquad\qquad\quad$ *- Temp. diff. (Δt - Dt) =* <u>*15°F.*</u>
$\qquad\qquad\qquad\qquad\qquad\qquad\qquad\qquad\quad$ *17°F.*

Suction pressure = 18.9 psig.

Refrigerant charge: <u>*200 Btu/min./ton*</u> = <u>*25.0 Btu/min.*</u> *= 8.0 oz (1/2 lb)*
$\qquad\qquad\qquad\quad$ *1/8 ton* $\qquad\qquad$ *50 Btu/lb (16 oz)*

Refrigerant was released and two (2) tubing connections were provided; one on the liquid line for the new filter-dryer installation, as well as an adapter to the unit's suction side, to provide a manifold connection.

The system was pressurized to check for leaks, followed by a deep vacuum process, then, the refrigerant charging process was initiated. When enough pressure was build, the new filter-drier was installed in the liquid line. Tested for leaks and proceed/finished with charging the refrigerant. The system worked without problem.

===

EXPERIENCE 11-B

Analysis

I had this experience with the mother of a former student of mine. While she was living in New York City, she received a refrigerator as a gift. Then, she moved to Florida and decides to take the refrigerator to that location. For several weeks, the refrigerator wasn't performing well. The student called me and asks me what he could do. Besides changing the refrigerator

I suggested putting on the back of the equipment a fan to blow air on the condenser's surface. It worked. The refrigerator performed much better.

Calculations

1. $Q = A (\Delta t.) \text{ "U" Factor.}$ 2. $A = \dfrac{Q}{(\Delta t) \text{ "U" Factor}}$

Where:

 $Q = Btu/lb$
 $A = Area (sq\ ft)$
 $\Delta t = Temperature\ Difference\ *\ (°F)$
"U" Factor = (Btu/sq ft/hr)

Note. The temperature difference (Δt - Dt) value to be applied, depends, on the type of cooling medium being used (Air - Water) as well as the ("U" Factor) value of the type of material used in the condenser's construction.

Example

The square feet (sq. ft.) area of the air-cooled, static type condenser, in a One Ton system, working at an ambient temperature of 80°F. (26.7°C.), with a temperature difference (Δt.) of 30°F. (-1.11°C.) and a "U" Factor value equal to 4 Btu/sq. ft. /°F. Should be:

$$A = \frac{Q}{(\Delta t.)\ "U"} = \frac{15,000}{30 \times 4} = \text{Approx. 125 sq. ft.} \quad (\text{1/3 H. P. 125/3} = \text{Approx. 42 sq. ft.})$$

===

EXPERIENCE 12-B

Diagnosis

I paid attention (as I, usually, do), to the system's operation and I couldn't hear in the evaporator neither hissing nor gurgling noises other than the running compressor. While the machine was working, I began to trace and feel the temperature of the refrigerant lines and it notice, immediately a crass connection error:

The compressor's discharge line (which seemed to me incredible) was higher than normal; it entered the condenser by the lower portion of it. (With that connection, it was like "raining upward.")

At this point, I did not have, to look for anything else.

I didn't even have to waste the time, of verifying the other lines (Suction and Liquid line) temperatures.

ALL the refrigerators presented the same problem.

Procedure

The repair consisted of inverting the condenser's lines connections. The change was performed, a service line was attached to the suction line, flare nuts were placed on the liquid line for the new filter-drier, and the parts where I worked were verified for refrigerant leaks.

After the vacuum procedure, refrigerant was added to up to 5 psig. A new liquid line filter-drier was connected and checked for leaks. The system was charged and the refrigerator worked normally.

Sheet metal was repaired, in all of the six refrigerators and finally, all of them were painted. Everything was OK.

Note: When I troubleshoot any refrigeration or air conditioning system, I usually use as many operating references as possible. Here, I try to list them:

1. Suction pressure
2. Discharge pressure
3. Start and Running amperages
4. Voltage values
5. Suction line temperature
6. Liquid line temperature
7. Evaporator's appearance (Dryness-wetness-etc.)
8. Liquid line's sight glass (bubbles, clearness, color, etc.) when used
9. Liquid control and evaporator's noises (hissing, gurgling, etc.)
10. Air flow through evaporator's and condensers. Where it is possible
11. Etc.

===

EXPERIENCE 13-B

<u>Analysis</u>

After walking inside the huge storing space, I decided to ask the person in charge if he had an idea as to where in the space was the meat damaged. I also asked the other workers the same question. They all gave me their idea, about the location.

<u>Procedure</u>

I directed myself to that area. Moved and touched several pieces of the hanging meat until I encountered some sites (pockets of "warm" air) where the temperature was a bit higher, than the rest of the space. For me, that was an indication of poor air circulation, caused by not enough space between the pieces of stored meat.

<u>Calculations</u>: (Ammonia R-717 - NH3) Normal readings.

Condensing temperature = Cooling Tower water temperature = 85°F.
+ Temp. diff. (Δt - Dt) = $\underline{25-35}$°F.
105–115°F.
Discharge pressure = 214.4–251.7 psig

Water temperature {In-Out Condenser (Δt.)} 10/12° = 95–97°F

(B. P.) Boiling temperature = Required (water) temperature = 15°F
$$- \text{Temp. diff. } (\Delta t - Dt) = \underline{-15°F}.$$
$$\text{Brine temp.} = 0°F.$$

$$\text{Brine temp.} = 0 \ °F.$$
$$- \text{Temp. diff. } (\Delta t. - Dt.) = \underline{-15 -25} \ °F$$
$$\text{Boiling point} = -15 -25 \ °F$$

Suction pressure = 1.25–6.15 psig (>=< -20°F)

Solution

Provide more space or gap between the pieces of meat, and make sure to move them more often to improve the air circulation.

===

EXPERIENCE 14-B

<u>Analysis</u>

Most of the times, when a seal/semi hermetic unit, (motor-compressor) is burned-out; it happens at the starting time. Taking into consideration, that the internal discharge valves on the valve's plate are closed, the flames-smoke-etc. produced by the burning motor does not have any other place to go, than through the suction line; since at that moment, this is the only compressor opening to the system.

The contamination; concentrates in the unit itself and the flames, smoke, etc. only travel a few feet through the suction line (depending on its size) toward the evaporator. Thinking on this, the cleaning process can be accomplished more easily and effectively performed.

On the other hand, when in a R & A/C system had been used incompatible substances, oil, refrigerant, tracers, etc. The contamination is present at all times and at a burning moment, not only the unit is in jeopardy but, the entire system as well; besides, the cleaning process is by far much more complicated, expensive and must of the times incomplete.

In any case, to accomplish the cleaning process, it is necessary to make use of other methods (changing oil/filter-dryers/etc.) in the liquid as well as in the suction lines; to minimize the contamination of the system, as well as the motor's fault again.

Then, as a precautionary measure (to neutralize the contaminant effect), the process of vacuum; at the end, and before charging oil/refrigerant, it has to carry out by a much longer time.

Today, it is not just to change the refrigerant (i.e. R-12 by R-134a. etc.), but also to consider that the oil may not be compatible; so, some other steps have to be performed for the change to be successfully. Think-Use common sense and consult with Manufactures.

===

EXPERIENCE 15-B

Troubleshooting is based on comparisons: Pressure, temperature, amperage, voltage and etc. readings, noises, (mechanical as well as electrical components) amounts of water, air, etc. Etc. actually, and the ones that the system is supposed to show-use; when it's working properly.

<u>Analysis</u>

The system operation analysis was performed in the following ways:

1. *Operating pressures were observed on the gauges on the board/wall.*

   a. *Discharge (260 psig.) and Suction (12 psig.).*

2. *Water temperature differences, sensed by thermometers and hand feeling.*

3. *Water returning to the cooling tower, was observed visually, and its temperature by the use of thermometers.*

   a. *Temperature difference (Δt - Dt) (25°F).*

   b. *Cooling tower, water level was normal; but water back was scarce and hot.*

4. *A pressure gauge was installed, at the vertical water line leaving the water pump.*

   a. *Water pressure (13 psig).*

Solution

After checking the system's entire operation (Notes 1-4 above) and the pump rotation (Electrical connections) I proceeded to stop and remove the water circulating pumps front cover, and found stuck there, (inside the impellers), knots of wood from the separators/ retarders of the water fall inside the cooling tower. The knots of wood were removed.

A new cooling tower water-pump set of gaskets was ordered.

When the new gaskets were installed, the system was tested and went back to normal operation.

According to the brine characteristics; the amount of brine to be moved by the pump, in this system; should be:

$$\frac{1,000}{W = (\Delta t.) \, Sh} = \frac{1,000}{12 \times .85} = \frac{1,000}{10.2} = \text{Approx. 98 gpm.}$$

==

According, to the refrigeration system characteristics; the amount of water to be moved by the pump, through the condenser; should be:

$$Q. = W (\Delta t - Dt) \, Sh$$

$$W = \frac{Q}{(\Delta t.) \, Sh} = \frac{267}{12 \times 1} = \frac{22.25}{7.43} = 2.99 \text{ gal/ton/min}$$

Where

Q = Btu's. /ton. /min. ($\Delta t.$) Temp. Diff. = 12°F Sh = 1 Btu/lb/°F 7.43 = lb/gal

GPM = 90 Tons x 2.99 = Approx. 269 gallons per minute

Note. For other operating calculations, see Experience # 13.

To have an idea of system's capacity, look at other calculations-references below.

1. Brine temperature = 3°F. (-16.1°C.)
2. Number of containers = 40 (6" x 18" x 30") = 3,240 / 1,728
                                               = 1.9 cu ft x 62.4
                                               = 117 lb (53 kg)
3. Total amount of water to be frozen = 117 x 40 = 4,680 lb (2,127 kg)
4. Brine specific heat = 0.85 Btu/lb
5. Total system capacity in Btu/hr (tons of refrigeration)

(1) Qs. = W (Δt.- Dt.) Sh. (water = 1 Btu's/lb)
       1,498 (60 – 32) 1
          = 1,498 x 28 x 1   = 41,933 Btu's.

(2) Ql. = W (LHF) (ice = 144 Btu's/lb)
       1,498 x 144   = 21,571 Btu's.

(3) Qs. = W (Δt.- Dt.) Sh. ice = 0.504 But's/lb)
       1,498 (32 – 18) 0.504

	°F		°C
	60	(1)	16.15
	32		32
	32	(2)	32
	32	(3)	32
	18		-7.7

Qt. = A. Qs. = 224,640 Btu's.
      B. Ql. = 673,920 Btu.
      C. Qs. = 33,022 Btu.
         Qt. = 931,582 Btu.
    + 15% = 139,737 Btu.
          1'071,319 Btu'/24 hrs.
          12,000 Btu. /Ton/Hr. = Approx. 90 tons.

==

313

EXPERIENCE 16-B

While, on the roof, I asked the mechanic, "Does the receiver have enough size, to store the entire refrigerant charge?" He did not answer me. I assumed that, or he did not understand what I said or that my question, for him, was not clear, important, or simply he ignored me. He probably knew that I was new in the company.

I followed him, we took the elevator and when we were going down, more or less, about the third floor, we felt a roar accompanied by a tremor; I look at the mechanic and he only shrunk his shoulders. We arrived at the first floor and we start making the repair.

While we were doing the repair, in a matter of minutes, the place was full of NYC Fire department men. They asked the mechanic some questions; I did not know the questions either his answers. We finished the repair and we went to the roof.

Regularly in an A/C system of this capacity, when it is of the self-contained type, the amount of refrigerant to be charge is easily calculated. See Calculations, below.

<u>Calculations.</u>

(*) R-22 Approx. charge of refrigerant = 200 Btu/min/ton / 86.7 Btu/lb
(latent heat value)
= 2.3 lb/ton/min x 5 tons
= 11.5 lb (+/- regular R-22 charge)

Although in this case we MUST take into consideration, that it was a big gap between the two refrigeration system components: The condensing unit (roof), and the evaporator (First floor) and that the lines (Liquid and

Suction had to be filled with refrigerant). The manufacturer does not know about this and because, he do not size the receiver to store the extra amount of refrigerant.

When we got back to the roof, the air conditioning's condensing unit components had almost disappeared.

==

EXPERIENCE 17-B

Product characteristics.

1. *Storage.*

   a. Crates size-content.

   L = 24" (*8 oranges*) x W = 12" (*4 oranges*) x H = 6" (*2 oranges*) -
   (*64*) Oranges 1/12" = 0.083 8/12" = 0.66

   b. Units/crates Approx. = 64 {64 x 8 oz (1/2 lb) = 34 lb}
   c. Number of crates   = 310 (10,500 lb/34 lb)
   d. Weight/crate   = 34 lb
   e. Number of shelves   = (310 crates/18 crates/shelve = 18 shelves)

2. *Product size:* Approx. Diameter:
           7.5 cm (3") x 8 = 24" (3") x 4 = 12" (3") x 2 = 6".

   Weight: Approx. 8.0 oz /each

Equipment characteristics.

1. Initial temperature   = 85°F.
2. Storage temperature = 32°F.
3. Specific heat value   = 0.90 Btu/lb
4. Relative humidity   = 65/70%
5. Water content   = 87.9%
6. Time of storage   = 6 weeks
7. Lights - ventilators   = 3 (40 w) (3 x 3.41 watts x 6 hrs = 3,672 Btu)
8. Workers/time   = 3 (approx. 1,500 Btu/hr) (6 hrs) = 27,000
                                              30,672 Btu x 6 hrs

Formulae/Calculations.

1. *Product.*

     Qs. = W (Δt/Dt) Sh
Where: Qs = Sensible heat (Total sensible heat in Btu)
     W = Weight (pounds)
     Δt = Temperature difference (°F.)
     Sh = Specific heat (Btu/lb)

Qs = W (Δt/Dt) Sh

   10,500 (85-32) 0.90

   10,500 x 53 x 0.90 = $\underline{500.850}$ Btu   = $\underline{41.74}$ tons = 2.3 tons/hour
                           12.000 Btu/ton   18 hours

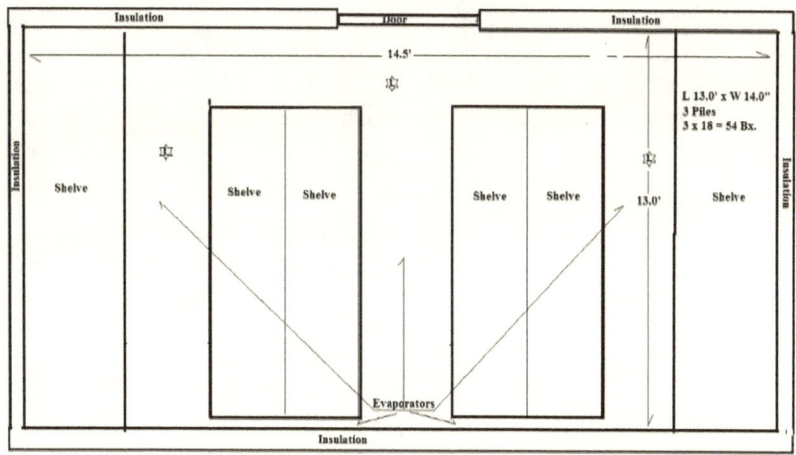

Room size (Inside the building).

317

Walls thickness = 8.0" (Hard board 1" x 2 = 2" + Fiber glass 6")

    a.   Inside walls -Ceiling-Floor. L = 14.5 x 11.0 = 159.5 x 2 = 319 sq. ft.
                                        W = 13.0 x 11.0 = 143.0 x 2 = 286 sq, ft.
                                        C = 14.5 x 13.0 = 188.5 sq. ft.
                                        F = 14.5 x 13.0 = 188.5 sq. ft.

    b.   Outside walls Ceiling.   L = 15.2 x 12.3 = 187.0 x 2 = 374 sq. ft.
                                        W = 14.3 x 12.3 = 176.0 x 2 = 252 sq. ft.
                                        C = 15.2 x 14.3 = 217.4 sq. ft.

Room dimensions: L = 14.5' W = 13.0' H = 11.0'

Room total space: = Shelves 18 (6 x 14" = 84/12"   = 7.0 ft.)
                     + Spaces (3) b/shelves   = 7.5 ft.
                                    Length = 14.5 ft.
                                    Wide = 13.0 ft.
                                    High = 11.0 ft.

Total heat load:

    1.   Product   = 27,960 Btu/hour.
    2.   Filtration/hour   = Outside walls:

                              Ends. 14.3' x 12.3' x 2 = 352.0 sq. ft.
                              Sides. 15.2' x 12.3' x 2 = 374.0 sq. ft.
                              Ceiling. 15.2' x 14.3'   = 217.0 sq. ft.
                                      Total area = 944.0 sq. ft.

Thermal conductivity.

Material characteristics: "K" factors (Btu/hr./sq. ft./1" thick/ 1°F.)*

a. Fiber glass   (6.00") = 0.25*   b. Hard board   (0.75") = 0.73*
c. Wood platform (1.00") = 0.80*   d. Cement mortar (3.00") = 0.50*

Formulae/Calculations: 2. Filtration

$$Qs = \frac{\text{"K"} \times A \times T \,(\Delta t - Dt)}{Tk.}$$

Where: Qs = Heat (total sensible heat in Btu)
         A = Area (square feet)
         T = Time (hr.)
    $\Delta t - Dt$ = Temperature difference (°F)
       Tk. = Material thickness (inches)

$$Qs = \frac{\text{"K"} \times A \times T \,(\Delta t - Dt)}{Tk.}$$

   a.   (Walls-ceiling outside. Hard board "K" = 0.73)

$$Qs = \frac{0.73 \,(K)\, 944'\text{ square} \times 1\text{ Hr.} \times 26.5°F.}{1.5"\text{ thickness}} = \frac{18,262}{1.5} = 12,175 \text{ Btu/Hr.}$$

   b.   (Walls. Fiber glass "K" = 0.25)

$$Qs = \frac{0.25 \,(K)\, 944'\text{ square} \times 1\text{ Hr.} \times 53°F.}{6"\text{ thickness}} = \frac{12,175}{6} = 2,029 \text{ Btu/Hr.}$$

   c.   (Floor. Cement mortar "K" = 0.50)

$$Qs = \frac{0.50 \,(K) \times 217.4'\text{ square} \times 1\text{ Hr.} \times 8°F.}{3.0"\text{ thickness}} = \frac{870}{3.0} = 290 \text{ Btu/Hr.}$$

d. (Floor. Wood platform "K" = 0.80)

$$Q_s = \underline{0.80} \text{ (K)} \underline{\text{ x 189' square x 1 Hr. x 8°F.}} = \underline{1,210} = 1,210 \text{ Btu/Hr.}$$
$$\qquad\qquad 1.0' \text{ thickness} \qquad\qquad 1.0$$

Qt. Filtration/hr. = (a) 12,175 + (b) 2,029 + (c) 290 + (d) + 1,210 =
$$\qquad\qquad\qquad\qquad\qquad\qquad\qquad\qquad 15,704 \text{ Btu. /Hr.}$$

3. Others (Handling/lighting/Etc.) = $\underline{30,672}$ Btu = 5,112 Btu/hr.
$$\qquad\qquad\qquad\qquad\qquad\qquad\qquad 6 \text{ hrs.}$$

Total heat to be removed:

    1. Product   27,600 Btu/Hr.
    2. Filtration 15,704 Btu/Hr.
    3. Other   5,112 Btu/Hr.
                   48,482 Btu/Hr.
         + 15% 7,272 Btu/Hr.
    Total: 55,754 Btu/Hr.   = Approx. 4.65 Tons
           12,000 Btu/Hr./Ton

Recommended Refrigeration system:

1. Five ton. Open or Semi hermetic type unit with an air-cooled condensing unit.
2. One or three gravity type evaporators (backroom or aisles). See diagram.
3. 230 Volts, Three phases, 60 Hz.
4. Other mechanical/electrical required components, according to system's characteristics.

Operating pressures.

Calculations: (R-12)

According to this refrigerant's characteristics, the operating pressures should be as follows:

a) Discharge pressure:

Condensing temperature = Ambient temperature = +/-85°F.
+ Temp. diff. (Δt - Dt) = <u>25–35°F.</u>
+/-110–120°F.

Discharge pressure = 136.1 to 157.3 psig.

b) Suction pressure:

(B. P.) Boiling temperature = Desired temperature = 32°F.
- Temp. diff. (Δt - Dt) = <u>15°F.</u>
17°F.

Suction pressure = 18.95 psig.

EXPERIENCE 18-B

Comment

Curiosity! It has never been a good advisor!

Sometime later, my wife's aunt asked me for the painting of its refrigerator. The painting job was carried out with the same white lacquer paint. Somehow I had the idea, that in the previous work it had not been letting the paint dried enough time. I bought the same type of blue gasket; this time, we let "dry" the painting by a longer time, and one entire week pass by.

I settle the gasket; took a precaution measurement, this time I put some type of powder between the door's paint and the gasket. We closed the door and even with the longer time and powder, there were the reaction and the same problem it happens again.

I was conceited and even today I think the same way, substances of unknown source or doubtful compatibility are not due to be mixed.

==

EXPERIENCE 19-B

Comment

There were no calculations, analyses, etc. for this experience, because it was only a procedure performed by myself. I did it that way because that was the way I learn to work where I'm from. Otherwise, I should use the waiting time cleaning the equipment so the customer doesn't see me wasting the time and his money. I guess.

Here, I remember a wise proverb "When in Rome, do as a Roman would do."

==

EXPERIENCE 20-B

<u>Analysis</u>

While the mechanic went to the van, I stopped the compressor and I conducted a more thorough inspection; first I saw, that the TXV had a "yellow" sticker; signaling that this system used R-12 instead of R-22 refrigerant. Then, I went to the back of the system, and I noticed in the equipment's ID. Plate the formula: CCL2F2 meaning: R-12 refrigerant.

When the mechanic came back and found the system off, he was very angry and he scolded me. I told him! Hold it! Then, I showed him the yellow color of the thermo expansion valve and the R-12 (Formula) on the ID, plate. He stopped reproaching me, went to the phone and talked to whomever it was that did the troubleshooting.

<u>Calculations</u>.

The amount of refrigerant charge needed in this A/C system should be calculated on the following manner.

(*) R-12 Approx. charge of refrigerant = <u>200</u> Btu/min/ton = 4 lb/ton/min x 5 tons
+/-50 Btu/lb (latent heat value)

= +/-22 lb

Then, with a little more confidence, I told him that one (30 lb) R-12 cylinder should be more than enough (*). He called the office and ordered it.

324

While waiting for the refrigerants arrival, he commanded me to bring a 50' x ¼" copper tubing roll. He closed the service suction valve and, connected it directly to it the tubing, extended the tubing so that the other end went out of the place and let the refrigerant escape into the atmosphere.

I thought! One "crime" after another!!

From me: No comment. The whole thing is self-explanatory.

This happened in 1968

I believe this information is more than effective today, with so many different oils and refrigerants in the market and in the trade.

==

EXPERIENCE 21-B

According to the last paragraph on part "A." We must direct our "investigation" on the mechanical parts or components of the equipment.

Analysis

I asked the mechanic, if I could see the internal thermostatic expansion valve component parts.

He showed me the various valve's components. When I had them in my hand I noticed, and let him to know, that the three (3) stems, located under the valve's diaphragm, which work is to push the valve to open; were flattened (shortened) at their end-points, this, due to the "Prolonged" work.

With this problem, the expansion valve NEVER feed the evaporator properly; creating the low suction pressure, the "Low boiling point" and of course the "Frozen evaporator."

Solution

I suggested calling the valve's manufacturer and asking for a new set of stems. In case them were not available (discontinued), then ask about the normal size of the stems and according to the information, to repair them. To accomplish this, I suggested using a "drop" of silver solder added to the stem's tips, to restore them to their original size' length.

At this moment, the mechanic came back and we left the place.

Several days later, I asked the dispatcher what have happened in that place. I found out the system was working well. I guessed they probably took into account my suggestion.

==

EXPERIENCE 22-B

<u>Analysis - Solution</u>

I looked at the motor's mounting and I suggested to the mechanic; that we could do the work without having to remove the motor from its site.

He show some indecision, but then he accepted my suggestion {I do not know (about the doubt)}, probably afraid of bending some sort of regulations) anyway, we perform the job without any major problems.

===

EXPERIENCE 23-B

Analysis

I checked the cooling tower operation and noticed, that the "eliminators" (whose purpose is to prevent (minimize) the water to escape with the air flow). These devices, that are located at the top of the cooling tower, between the water "spreaders" and the blowers, were missing; and that was the source of the problem.

I suggested the mechanic to call the office and give them the information (Tower's design, manufacturer, model and serial numbers, eliminator's size, etc.), so they can be ordered and then sent someone to install them. So, he did it.

==

EXPERIENCE 24-B

<u>Procedure</u>

1. *Two separated manifolds were connected to the system's suction/
    discharge packing gland service valves; and to the oil's pump
    discharge connection; the contaminated refrigerant was released to
    the atmosphere thru:*

    a. *The hose connected to the system's suction side service valve,*
    b. *By opening the # 1 manifold's left side valve,*
    c. *Then, let the refrigerant go, thru the manifold's center hose to
        the atmosphere, thru a 1/4" tubing extended to outside the room.*

    I believed, the contamination was concentrated in the compressor . . .

2. *The system was decontaminated by introducing nitrogen:*

    a. *Through the # 2 manifold center hose,*
    b. *Opening the manifold's right valve,*
    c. *To the high side manifold hose, which was connected to the
        discharge-packing gland compressor's service valve and*
    d. *Let the gas escape through the center hose of the # 1 manifold
        connected to the low side.*

3. *Then, we proceed to remove the old compressor, carefully and by
    using safer and appropriated procedure.*
4. *The new compressor was installed.*
5. *In the existing liquid line filter (flare nuts) location, we connected
    a piece of tube the same filter size length. The same procedure was
    performed at the location of the suction line filter.*

We performed all the procedure in the appropriate form (besides the two pressure gauges sets, an ammeter, a voltmeter, etc. were used) and at about 2:45 p.m. after carrying out a good deep vacuum process, through the two manifolds; we added refrigerant through the low pressure side till getting and keep a positive pressure of about 5 psig. While the refrigerant was leaking, new suction as well as liquid line filter/dryers were connected.

After checking for leaks, we continued charging refrigerant in a liquid state, through the system's liquid line till the pressures (cylinder, system) equalized (about 140 pounds*); the manifold's connection to the king valve was closed, and the equipment was started. We continued the vapor refrigerant charge through the low side.

* Total charge was approximately 225 lb.

The oil pressure was also verified during the entire system's operating process.

The complete operating process was completed. At about 4.30 when were almost finish, the mechanic began to carry to the truck all the tools that we had used. In the meantime, I remain verifying the system's operation.

After removing the rest of tools and thinking everything was OK, we were leaving the place; when we almost reached the exit's door, we hear a strange noise coming from the motor that we had just changed; I ran back to the room and disconnected the power.

Then, I connected the tools to the system and within few minutes; started it again. Everything was normal. Nevertheless, I suggested to the mechanic to ask to the person in charge if in the last 10 or 15 minutes, they had used other machine; different to which they were using normally.

The mechanic was tired, but a little reluctantly, he formulate the question; we found out by that person several things:

1. They had worked a selected cards machine.

   This machine, according to the person in charge, work sporadically, it was only used once or twice per week.

2. Since the installation of that machine, the previous year, this had been the third time that took place the burned out of this compressor.
3. By the investigation that we conducted, we found out that in the same electrical circuit of the A/C system it had been connected the cards selecting machine.

This machine had the following characteristics:

a) A single circuit breaker connected to one of the 440 volts A/C's power supply and
b) A neutral line to get 220 volts. 60 Hz.

Then, I suggested to the mechanic to ask the person in charge if they can run the machine once again.

He agreed; when the machine started to work, the compressor presented/displayed the noise-problem again, then, quickly, I took the voltage readings, in two (2) of the three power supply lines, (its value it decreased from 440 to approximately 300 volts, whereas reading of amperage rose very over the normal values). As fast as possible I de-energized the system. Several minutes passed and after the selected machine was de-energized, I start the A/C system again everything was normal.

Then, I suggested the mechanic to call the company's office and let them know, about this problem. In the office, they did not pay that much attention and they decided that we left the system working. By my insinuation, a few days later I asked to the mechanic, if he could find out what had happened with the compressor. He did it, and we learned that the compressor had been burned out again.

This time the dispatcher said to the mechanic to be aware that, in the near future (maybe a couple of days), the compressor's manufacturer was sending a supervisor to meet with us. That he would let us know when.

By then, I was pending and somehow nervous. I wanted to know what was going to happen when the compressor's manufacturer supervisor came. And *the day arrived; we were there before this person showed up. We coordinated what the mechanic was going to say. "Lubrication."*

When the man came, he greeted us, then, he asked the mechanic "What do you think was the problem? The mechanic answered to him "Lubrication" and . . . that was it. Then, he went away.

I'm sorry! But I had a big laugh of the compressor manufacturer's representative procedure. Furthermore, of the entire process.

==

EXPERIENCE 25-B

With so many R-12 empty cylinders. I thought, I do not think this system has any refrigerant leaks, if it it's any refrigerant leak in it, in a so reduced space, as soon as I started the torch (or any leak detector device) it will show the leak right away. Just for the fun of it, I lighted the torch and the flame was of a blue color; indication that "there were NO refrigerant leaks."

<u>Analysis</u>

Then, I thought the refrigerant could be someplace. And I got myself to find it.

By using a service wrench, I opened the packing gland service valve type (king) of the old condenser outlet and felt/hear the flow of the refrigerant leaving from it, the pressure it began to raise on the wall's low-pressure gauge and the machine started.

This time I assumed, that the valve coming from the compressor, in spite of being closed, front seated (to the old condenser), it had a leak allowing the refrigerant to continued flowing, and letting it be stored there.

Let me explain: *The temperature in this space was about 90°F. With this temperature, R-12 inside the old condenser had a pressure of about 99.6 psig. With the water from the cooling tower at a temperature of 80°F. (Outside temperature) the discharge pressure should be somewhere between 125 (105°F.) and 170 psig. (155°F.) So, with the machine in operation we had a pressure difference of at least 30 or 40 psig. (+/-45°F), the refrigerant easily flows, through the leaking valve, from the compressor to the old condenser.*

A while later (about 30 minutes), the mechanic returned and he, all excited asked me what I had done; that the equipment was working and the temperature in the restaurant was beginning to descend.

The best I could, tried to explain to the mechanic my theory, about the refrigerant being held in the old condenser; he said that the old condenser had been replaced because it had a "refrigerant leak."

I verify (feel it by hand) the temperature difference between the condenser's water supply-return lines; it was about 14°F.

I also saw on the wall's discharge pressure gauge,. and notice that (although the water thru the condenser was running) it was too high (+/-195 psig). I also checked the liquid line temperature and found it to be higher than normal (about 110°F). These facts gave me the idea of an excess of refrigerant.

I proceed to remove the excess of refrigerant, by using the empty tanks that were in the place.

To accomplish this procedure, one by one, I pull a vacuum process in each of them as I was recovering the refrigerant through the king valve at the condenser's outlet. I was able to remove at least 150 pounds excess of refrigerant.

Although, I believed that was nothing wrong with the old condenser, the office ordered to cancel it, anyway; it took us a few hours, to cancel completely the line from the compressor to the old condenser.

===

EXPERIENCE 26-B

<u>Analysis</u>

After verifying that was nothing wrong with the water supply and getting more familiar with the water system, it was assumed, that something maintained the check valve supposedly located in the water pump's piping outlet, in the open position; this way when the pump stopped, all the water in the pipes returned to the tower flooding the surrounding areas.

The A/C system was stopped; then, we proceed to drain the cooling tower, until at stopping the water pump; the water level was no dangerous of flooding.

The only problem that we had was, to find a pipe wrench, big enough (24") to unscrew the valve/pipe and to check/remove whatever was creating the problem.

Solution

After talking to the dispatcher at the office, we were able to get the pipe wrench. Once the check valve at the pump's exit was disconnected, a big piece of rubber was found there; that piece remained stuck in the valve's seat and keeps it in the open position.

The problem was removed and the system went back into normal operation.

==

EXPERIENCE 27-B

Analysis

I assumed that in some place of the building's air supply ductwork, may be a sensor (flow switch) that "it felt," when the supplied air flow from the building were interrupted, and closed another electrical circuit to the control circuit of the computer's room A/C equipment. No matter how hard we look, and traced the electrical installation to the old control, nothing could be found.

Solution

I did not wanted to waste any more time. I proposed the installation of an airflow sensor. The same was acquired it and it was located in the supply air duct; the electrical connection was performed. Once its installation was completed, the equipment was back to work without problem.

===

EXPERIENCE 28-B

<u>Procedure</u>

Working properly and honestly, without trying to prove anything, the entire work was completed in one week (approximately 40 hours). Working hours 8:00 a.m. to 12:00 noon and 1:00 p.m. to 5:00 p.m.

The day I called to report the job's completion, the dispatcher was very surprised.

Several weeks later (according to my "spy" in the office), I knew that the dispatcher had called to the serviced building, to verify the operation of the equipment. And when he knew that all systems were working perfectly, they called to the office, to the mechanic that did the same service in the previous year, and asked him; why he had spent almost a month to complete this same work; and in the following week they had received at least ten service calls.

What he answered was that he had "nothing to prove."

===

EXPERIENCE 29-B

Comment.

To prevent a problem like this, it is a very good idea to know how to read/ interpret an electrical diagram; not only that, but to get familiar with the Manufacturers information, usually printed on the device's body, as well as in the equipment diagrams.

Thermal overloads of this type; have (clearly marked) two electrical connections: One, for the control circuit (Low voltage) of 24 volts and the other (High voltage) for the line circuit 230 volts.

The overloads damage was caused by the wrong electrical connection: Line voltage was applied to the low voltage electrical circuit. The high current passing through their contacts caused the damage.

Sometimes, a fault like this can be seen by the color of the device's contacts; if it is not possible to see them, an ohmmeter can be used to read the normal "continuity" that they should show when they are in good condition. In this test, any "resistance" or "openness" reading showed by the ohmmeter should be an indication that they are faulty.

===

EXPERIENCE 30-B

While he was explaining to me the whole problem, and because the basement ceiling was too low, I was seated on top of an empty bunch of potato sacks.

All of the sudden I felt heat and vibration under my seat, I remove the sacks and the compressor of the service needed system was "discovered" under them.

The manifold gauges were installed in this compressor and found the system was cycling by the low-pressure control, due to a lack of refrigerant. Luckily, the refrigerant leak was located right there at the compressor's suction service valve's flare nut. The flaring connection was retightened and refrigerant was added. The entire system operation was checked.

The whole thing took about 30 minutes to repair it.

I recommended designing a plan like, where all the system's characteristics were listed and identified. As well as marking every basement's machine with a number/letter matching each machine on the store's first floor.

==

EXPERIENCE 31-B

<u>Analysis - Solution</u>

I begin, by looking for a usual tool used in cases like this, a ball-bearing/ impeller extractor. Since there was none in the place; I put some oil on the shaft, as well as in the impeller's screw holder hole. Then, by using a screwdriver, I pried at the impeller away from the pump's body, and at the same time with a rubber mallet, I struck the axis end inward. To the third blow of the mallet, the impeller came out.

I seated to wait for Rodriguez to return. When he came back, he asked me what I did it to remove the impeller. I explained to him what I had done and I left.

===

EXPERIENCE 32-B

Particularly I always believe, like in everything else, sharing knowledge help to widen our thoughts and by doing so, work is better performed.

I had the same experience with some other workers. At least, I had the satisfaction of someone getting the benefit of my knowledge.

===

EXPERIENCE 33-B

<u>Analysis</u>

Note. Experience may show us, that as many as twenty (20) detailed reasons, can give us answers for this question.

Remember the question # 2 asked in Part "A"? Let's give it the try.

Compare your list, with the one given at the end of this experience, on page # 346.

Once on the roof, and gotten the unit exposed, *I bent myself, to watch the oil level in the crankcase's sight glass. I didn't see any oil's level; (I assumed: there was no oil or too much of it); by using a lantern to see better inside, I noticed some "bubbles" that rose from the bottom of the crankcase. This indicated to me of an excess of oil in the unit*

I followed the refrigerant lines along the roof's floor and about 7 feet from the equipment, found a sight glass liquid indicator which I un-covered it, it was darker than usual. For me, meaning a lot of oil mixed with the liquid refrigerant.

I stopped the condensing unit and returned to the computer's room.

To melt the ice covering the evaporator, I left only working the evaporator's fan motor.

Then, I spoke to the person in charge; I asked him how long he had worked there? He looked at me, seems it like annoying or disturbed; I ask him for an excuse and explained to him that I simply wanted to know if at any time he had seen this system working well.

He understood me and answered that he had but of 8 years of work there; and yes, he had seen the system working well, but that had happened a long time ago (about 6 or 7 years).

In order to keep the service's control, the service company it used green fine card-boards of 8.5" x 11" size, 30 lines per side; in them, it was printed the date, reason for the service call and headings indicating: Actions applied to "correct" the problem.

In a place of the room, near the evaporator's location, I found a bunch of about thirty of these fine green pages. I got myself to review them. I went faster until arriving more or less at seven years back. Bingo!

In one of them, *I found that at that time somebody had changed the compressor; and according to what it was written, because "It did not pump." I studied the case I reached the conclusion for the compressor's mechanical fault, it may had been caused by (One of many reasons), among others liquid refrigerant returning to it.*

I observed the very low setting of the thermostat on the wall, which may produce the refrigerant-oil mixture compression; therefore, causing the compressor's damage.

Then according to the analysis, when the damaged compressor was detached, the mechanic "forgot" to remove the oil from the system. (New replacement compressors bring its complete oil charge.) Now the system-compressor had an excess of oil.

Since I believe, the cyclical compressor's operation does not allow excesses of oil in the crankcase; thus, the oil goes out; it does not stays in the condenser "Too hot in there" so, it lodges in the evaporator, interfering with the normal refrigerant boiling; by occupying the space and producing this way, a low operating suction pressure, so low that the boiling point goes below the freezing point of the moisture in the air (Dew Point Temperature); therefore freezing up the water in the evaporator's surface.

Task - Solution

The excess of oil must be removed, in order for the system to return to its normal operation.

<u>Procedure</u>

I went down to the truck and brought up a 30 lb cylinder of R-22, the pressure gauges set, service wrenches, etc. When I returned to the computers room, still some ice was on the evaporator's surface. I went to the roof, appropriately connected the manifold's gauges, and found an empty container (a gallon of paint) that I used to collect the oil that I was going to extract from the system.

I went down to the computer's room and observed that some of the ice was still covering the evaporator's surface. I returned to the roof started the system and through the compressor's discharge service valve, the manifold right side valve and center hose; little by little I began to let the vapor refrigerant to leave; a great amount of oil, as I expected, came out as well. I had to add refrigerant several times to the system; to replace the one that I left get out with the oil.

After about twenty-five minutes of performing this operation. I began to see the oil level in the compressor's crankcase sight glass. At this point, I went down to the computers' room, to verify that all the ice had disappeared from the evaporator's surface. Since there was only an small portion of ice on it, I got back to the roof and continued the procedure until the oil level reached the proper point, about the middle of the oil-sight glass, then, I had managed to remove approximately ¾ of gallon of oil; let the system run and verify its entire system's operation:

Pressures, amperage /etc. and I finish by removing the service manifold: Opened widely and momentarily both manifold's valves to return the oil-refrigerant mixture from the high side hose into the system. Once I accomplished that, the manifold's valves were closed and quickly disconnected the high side manifold's hose, from the discharge service connection was quickly disconnected.

Then, disconnected from the systems the low-pressure side manifolds hose. Put the covers back.

When I got back to the computers room (approximately 9:45 a.m.) the place was getting cold and the worker congratulated me. Then, I request to him permission to use the telephone, to call back to the office; he "denied it," instead he asked me to let him do it first, that he wanted to speak with a company's supervisor.

==

Reasons for an A/C frozen evaporator:

A. *Air flow*:   a. Dirty filters   b. Dirty evaporator

B. *Electric motor*: a. Loose electrical   b. Lack of lubrication
                        connections
                     c. Wrong electrical   d. Bearings-Friction sleeves
                        connections

C. a) *Blower*:   a. Dirty   b. Loose   c. Wrong rotation

   b) *Blades*:   a. Wrong pitch   b. Number of blades
                     c. Diameter   d. Loose
                     e. Wrong rotation

D. *Refrigerant*: a. Shortage   b. Restricted flow

E. *Thermostatic expansion valve*:

                     a. Misadjusted   b. Dirty filter
                     c. Wax deposits   d. Bulb (loose charge)
                     e. Bulb location   f. Wrong valve
                     g. Partially stuck closed

F. *Oil*:   a. Excess   b. Wrong

G. *Controls*: (Misadjusted, wrong, defective, etc.)
        a. Thermostat.   b. Limit.   c. Low pressure.
        d. Water regulating valve.

H. *Other*: (cooling medium)

        a. Too cold (ambient conditions)

I. Others.

EXPERIENCE 34-B

Not much from the "information" received from the engineer; but, from the one obtained when several of the A/C systems water supply-return lines were checked; I thought it was not necessary to verify any other systems.

Diagnosis

These water supply/return temperature's differentials, demonstrated to me that the problem was not in the A/C system's condensers themselves; but in the temperature/amount, of cooling water coming from the cooling tower.

Note. According to New York City climatic conditions, the normal lower temperature of the water from the cooling tower in summer is about 78°F.

I returned to the mentioned "Engineer" office, and I asked him where the cooling tower was located; he did not know and it had to call to somebody else to finding it out.

When they informed me the cooling tower location: Street level, around the corner; I went down there; enter the place and saw an enclosed type-cooling tower.

On the side of the cooling tower's structure, was and opening/inspection window like. I opened and was able to see the water level; it was normal, but the returning water falling into it, was not only very scarce, but unusually hot. The water inside the cooling tower was at about ambient temperature. At the time it was about 85°F.

Next, I verified the water's pump rotation; it was right. By using a pressure gauge I measured the pump's water discharge pressure, and I found it to be too low (approximately 15 pounds).

I went back to the engineer's office and told him that I had to stop the entire system to verify the water supply, as well as the return conditions. Arrogantly, he refused to let me stop the system; instead, he threatened me with calling the company and a make a complaint to a supervisor. I said to him that that was fine with me.

He took the telephone and made the call; he asked for the supervisor, and when he was at the phone, his complaint was that I wanted to stop the system, and he would not let that happen. All this time, I was standing in front of this guy's desk.

What followed next, as somebody at the office told me later, was that the supervisor asked the dispatcher who was in that service call. He answered him, "José." Then the supervisor told the engineer, "Listen to me, that mechanic who you have there is not one of the best ones but the best mechanic that we have in this company. If he is not good enough to serve to you, then nobody will be able to do the work that is needed."

The previously mentioned engineer looked at me and a little reluctantly said, "It's OK, do whatever you believe is necessary."

I went down to the cooling tower once again, stopped the system, and isolated the water pump by closing the valves at its sides (in and out). Then I removed from the water pump's the front cover and found that, the impeller's spaces were completely filled with wood knots coming, I guessed, from the wood separator planks inside of the cooling tower, and for that reason the pump did not move enough water through the systems for their normal operation.

As it was already almost five o'clock in the afternoon, and I had to go to the school, I called to the office and asked to the dispatcher to send two workers so that they could clean the cooling tower and the water system in general.

I also ordered a new packing-gasket set for the water pump because the old one was broken when I removed the water pump's cover. On the following day, I knew that both workers had been there till three o'clock in the morning.

The operation of the system, returned to normal three days later, when the water pump gasket packing arrived, and somebody else was sent to change it.

==

EXPERIENCE 35-B

In a system with these characteristics, at the outlet of the room operating at a higher temperature; a mechanical device called (E. P. R.) "Evaporator Pressure Regulator" is used to keep the refrigerant boiling point value, within the room's specific conditions.

Analysis

After verifying the operating system's pressures, we found them to be OK. But when installing a gauge at the room 2 evaporator's outlet, we found that the evaporator pressure regulator was out of adjustment.

Calculations (R-502):

According to the characteristics of this refrigerant, the operating and adjusting pressures should be as follows:

a) Discharge pressure:

Condensing temperature = Ambient temperature = $\quad$ 80°F.
$\qquad\qquad\qquad\qquad$ + Temp. diff. (Δt – Dt) = <u>10–12°F.</u>
$\qquad\qquad\qquad\qquad\qquad\qquad\qquad\qquad\qquad$ 90–92°F.

Discharge pressure = 185 to 191 psig.

b) The normal BP for -15°F room number 1 (Δt/Dt = 15°) should be:

Desired temperature   = -15
- Temp. diff. (Δt - Dt) = – <u>15</u>°F.
                = -30°F.
BP for R-502   = - 30°F   = 8.7 psig

Suction pressure = 8.7 psig

c) The EPR adjustment should be the following:

The temperature differential (Δt - Dt) for medium temperature application, it could be equal to 10 or 12°F.

Desired temperature   = 38 °F.
- Temp. diff. (Δt - Dt) = -<u>10</u>°F.
                = 28°F.
BP for R-502   = 28°F.

Solution

This pressure was taken on the connection located at the EP regulator. It proceeded to adjust it to maintain the pressure corresponding to 28°F (61 psig).

The entire system operation, as well as the control's adjustment, was verified.

===

EXPERIENCE 36-B

<u>Analysis</u>

The compressor was running because the low-pressure control was found defective.

For the "ground" ohmmeter's reading effect; I suspected the presence of water inside the compressor.

The compressor was disconnected; when it turned it upside down to drain the oil, along with it, it came out at least a half gallon of water. It assumed the water came from the evaporator's dripping tray; went inside the compressor, and was "producing" the passing of a current effect.

How the water from the dripping tray got into the compressor?

According to the realized study, the evaporator dripping's tray drain was partially clogged; that explains the presence of water in the tray. In this specific case, with a refrigerant leak in that location if the system kept working, once the refrigerant finished leaking, the vacuum created in the system by such operation, sucked the water (deposited there) into the compressor.

<u>Procedure</u>

I thought that by blowing the system with nitrogen then simultaneously energizing the crankcase resistance and pulling a good deep vacuum procedure, the problem could be corrected. I called the company and I asked the dispatcher for a cylinder of nitrogen with its pressure gauge/regulator, a new 5/8" filter-dryer, a gallon of oil Capella "D" SSU 150 and a new dual-pressure control. One supervisor, wanted to know what the problem was and I told him of the rupture of the suction tube and the water in the system.

Water? He sounded surprised. "The system is air-cooled" he said. I try it to explain to him but I guessed, he did not understand. Anyway, they sent me the material that I requested.

After finishing draining the oil, the compressor was put back in its place, the discharge line connected to it as well as a piece of 1/4" by 3" length copper tube to the same discharge line. A connection (of a 6" long, 5/8" piece of tubing with flare-nuts) was connected on the liquid line filter-drier's location as well.

The manifold's high side hose was connected, to the added service discharge line piece of 1/4" copper tube.

When the requested materials arrived:

a) *Set the nitrogen pressure regulator, down to 150 lb then, through the manifold's center hose and the 1/4" tube on a discharge line; blew the system, letting the gas out through the still disconnected compressors suction line tubing connection (sporadically covering the opening) to pressurize the system. This procedure was repeated, till I thought the system was sufficiently "dry."*
b) *Attached the suction line to the compressor.*
c) *Added 5 lb pressure R-22 and check for leaks.*
d) *By using nitrogen, increased the system's pressure step by step till getting up to 150 psig. No leaks were found.*
e) *Pressure was released and a deep vacuum was pulled from the compressor both sides, for about 75 minutes.*
f) *During the vacuum process, the crankcase heater was energized.*
g) *By using the vacuum in the system, the oil charge was added.*
h) *Add R-22 until the pressure rose to a positive pressure. While the refrigerant was leaking, the new filter-drier was installed on the liquid line.*
i) *Check for leaks on the filter-dryer.*

j)   *Proceed with charging the R-22 by weight (approximately 15 lb of refrigerant) until getting the calculated operating pressures. As well as the normal liquid and suction line temperatures.*

The dual pressure control was changed and set for the system's normal operating pressures.

The job was finished, by verifying its entire operation. When the job was almost completed (at a little past 3:30) a couple of supervisors came to find out about the repair. I showed them the water. They admired not only the work but also the short time used to perform it.

===

EXPERIENCE 37-B

I think my reaction was the right one. I must not; supposed to get into any discussion with anybody, at the working place.

And I thought, the best way to prevent problems like this; the next time I should ask to the person in charge, if I can do the maintenance service.

That's all the comments I had for a situation like this.

==

EXPERIENCE 38-B

As in other experiences, I'm not feeling like wasting your time or mine. So do/make the comment or comments that you think are appropriate. There will always be people like this. And I think there is nothing that we can do to change them.

===

EXPERIENCE 39-B

<u>Analysis</u>

The procedure to exchange the controls connections was analyzed, and it was found that it was not possible do it easily. To change the high-pressure control's connection accordingly to where it was connected, it was necessary to remove the entire refrigerant charge from the system.

The mechanic, who worked here before, let the entire refrigerant charge go to the atmosphere. What a crime! and waste of money! I was not going to do the same mistake.

I took the necessary action (and just for safety, jumped the controls) short-circuited them, properly installed the service manifold gauges to the system, and checked the system's operation, verifying that there were no other problems.

With all the information in mind, the first decision I made was of changing the dual pressure control. Therefore, I called the company's Office and told the dispatcher my findings; then, I requested to him a new dual pressure control. In addition, I asked if he could get for me, ten (10) R-12 empty cylinders of 125 pounds capacity each.

As almost always that they sent me to make some job, this time he was strange of my petition but, at this point they already knew me, and without any more questions he said, that he was going to try to obtain the cylinders, he would let me know as soon as possible.

In the meantime, I went to the superintendent's office and ask him if he could provide me with some ice. He wondered how much I needed and I told him the greater amount possible; I also requested to him, if he could please help me; by sending the ice to the machine's room or compressor's location. This man, lend me a great aid, because he also provided two workers to help me. I brought and prepared the vacuum pump. When the ice came (I guessed about two hundred pounds) I ask them to pile it up on the floor.

I place the controls in their place, connected the Low-pressure control and put the High-pressure control connecting tubing; ready to be connected.

About thirty minutes later, the dispatcher called, he told me that the cylinders were on my way. When the cylinders arrived, they helped me to put them on the floor above the ice to cool them. I did pull a vacuum process, to each cylinder.

Then, I removed the connected manifold high side hose from the discharge compressor's service valve and connected to the "king" valve at the condenser-receiver's outlet.

The system was put in operation; the king valve was closed (front seated), for a system's pumped down procedure; then, started removing the system's liquid refrigerant into the cylinders.

In order to know the amount of refrigerant in each cylinder, we raised it up in one end, valve's side and by doing so, we sensed more or less wise the weight and how much refrigerant was inside, in each of the cylinders. Thus, we were doing it, cylinder by cylinder until we recovered; almost the entire system's refrigerant charge.

When the high side pressure gauge showed a low enough reading (a few pounds), quickly stopped the system. Disconnected the previous discharge pressure control's tubing and connected the new one (I lost but a minimum amount of vapor refrigerant). After I finished connecting it, a check for leaks was performed,

The system was started and the recharge of refrigerant was initiated. The time was about 3:30 in the afternoon.

The equipment was already in full operation, when two company's supervisors appeared in the place "to look around"; the work that was been performing there. They were admired not only that we had recovered the entire refrigerant charge; but, completed the work in so just such short time.

===

EXPERIENCE 40-B

Analysis

Sounds un-believable that a piece of metal, of about 2 1/2" thick, breaks so easy. I thought, may be a defective material . . . But, no way. Three times?

This problem, it makes me curious and I wanted to know what had happened here.

I performed the following procedure:

Calculations: (R-12)

According to the characteristics of this refrigerant, the operating pressures should be as:

a) *Discharge pressure:*

Condensing temperature = City water temperature = 60°F.

Temperature difference through the condenser: +/-25–35°F.

Discharge pressure = Approx. 108 to 125 psig.

b) *Suction pressure:*

This calculation should be:

Desired temperature = 32°F. - 15°F. = 17°F.
B. P. for R-12   = 17°F   = *19* psig

Suction pressure = +/-19 psig.

The service gauges were connected to the system, and the operating pressures were verified by a pair of minutes. Early in the operation, I encounter the suction operating pressure was too low (R-12 Suction Pressure = 8 psig. B. P. = - 2°F.), and for me, this was the reason for the problem.

According to these operating pressures, the machine produces "a very cold" ice. An ice so hard, that the scraper it cannot remove it from the evaporator's surface, and breaks.

It was not the first time. Often, the clients express its wish of a "harder colder ice," trying to obtain from the ice to "remain by a longer time"; the service technician trying to please him does adjustments in the system's operation.

As well as "nothing is perfect" thus, also, everything has its limit.

I spoke to the owner and I explained to him my theory. He thought it for a moment and it gave me the reason. He requests that please, when the spare part comes; make sure that was me who return to make the work.

===

EXPERIENCE 41-B

Comment.

To solve the present problem, in our benefit we had to remember: that at spring time; we were trying to prepare the system for the summer's operation. It is always advisable and economic (if possible), to remember the important features of every job we attend.

Analysis

By checking on Experience 37, we may remember that this A/C system had a compressor with a five tons capacity, uses R-22 and it had a remote air cooled condenser. At that time I had to leave the job incomplete. So, at this time, I may continue where I was interrupted.

Procedure

The front A/C system's covers were removed, to gain access to the semi hermetic unit.

The manifold gauges were installed and notice:

1. *The discharge pressure was higher than normal (about 260 psig.).*
2. *The suction pressure was higher than normal as well (about 80 psig.).*
3. *The liquid line temperature was higher than normal (by tact +/-95°F).*

No need to check for anything else.

I went to the condenser's location. It was the first time I got here; I found the condenser's surface dirty (a mix of grease and dust), I guess for being near to the kitchen air exhaust.

Solution

The condenser had to be washed and cleaned by using water, a brush and some dissolvent that I got with a restaurant's employee.

Without wasting any time, this work took me a little more than 2 hours. The person/owner or whatever is very diligent and complacent today. He even offers me lunch, but I refuse it.

I finish my work; by verifying the entire system's operation. I get the customer's approval signature; he gave me a $50.00 tip. I only it would hope, that people were but little comprehensible with the people who, in some way, serve them.

===

EXPERIENCE 42-B

Analysis

Once I arrived in the place, I went to the second floor and study the problem; finding out that somebody (some time ago) had removed the eliminators that are supposed to be located on top of the cooling tower; between the water sprays and the (blowers) moving the air counter-flow through the cooling tower.

The function of these devices, as its name implies, is to separate, the water from the air stream; limiting the exit of the water in steam form, from the cooling tower and trapped it back to the tower. Thus, the problem had been produced by the absence of them. (See also: Rooftop "Flooding roof" Experience 23)

Solution

The pertinent information was taking: Tower's manufacturer, model and serial number, dimensions/size of the eliminators, etc. and it was reported to the dispatcher/supervisor.

Time for the completion of this work; about 25 minutes.

A few days after this service call; the dispatcher lets me knew about the eliminator's arrival and their installation had been performed and with it, the end of that problem.

===

EXPERIENCE 43-B

<u>Analysis/Procedure</u>

Once I was alone, front of the equipment, I set the equipment's thermostat in the "Off" position. On the wall, near the equipment I raised (connected "ON") the lever of the electric power supply box, where a fuse of the plug type was located; and by using the voltmeter, took the voltage reading from the fuse's (top) inlet and the neutral line; and I got a reading of 115 volts.

Then, I proceed to take the voltage reading from the fuse's (bottom) outlet and the neutral line. The voltmeter registers merely 65 volts.

Lowered the lever of the electric power supply box; fitted the analog meter in ohms scale x 1 Ω, verify the "ON" position in the thermostat and took the reading in the connecting wires that went to the equipment; the meter registered, to me, a normal reading of 2.5 ohms.

After this test, I set the ohmmeter to the x MΩ (100,000 ohms) scale and I submitted the electric circuit to a "ground" test (One Ohm's meter prong terminal to each power supply lines and the other meter's prong terminal to ground), this gave me a negative (Infinite) reading.

I went to the truck and I got a fuse plug type, 30 amperes fuse. When I returned to the cooler, I watched that nobody was around; and changed the defected fuse. After verifying the power supply voltage, I initiated the equipment's operation, and verify the starting (LRA), as well as the running (FLA) amperages.

The equipment worked perfectly: suction and liquid line's temperatures, etc. I removed the fuse, and I went out of the restaurant to call the company. I informed the dispatcher of what I had found, and I asked him what I had to do. He had me waiting on the telephone while he spoke to a supervisor.

A few seconds had passed, then, the dispatcher told me to change the equipment. Thus, I went to the truck; brought the unit in, change it and performed the rest of the procedure until the system was working perfectly.

===

EXPERIENCE 44-B

<u>Analysis</u>

Since the problem was an electric one; and I knew the location of every component of this A/C system. I went to the electric motor's location at the attic. Using the appropriate electric tools (Voltmeter and Ohmmeter) I reviewed the unit's electrical circuit and found the motor-compressor in 2 phases, in addition it was "grounded" that was it, it had to be changed it.

Thus, I let it know to the "client." Who still, almost pleaded me that "do something." But . . . There was nothing that I could do, but, to order a unit's replacement.

I called the office, reported the problem and give the compressor's characteristics. So they can order the replacement.

On more time! There are no more comments for this experience.

===

EXPERIENCE 45-B

<u>Procedure</u>

I'm not very customized to wait, and begin to think the form to exchange the compressors.

I found two pieces of metal angles and I placed them in oblique form, so the compressor's can slide on them. Taking down the old one, was no problem. To raise the replacement, on the platform, it was not an easy task. I began to push raising the compressor; when it was more or less in the middle of the route, I felt a tear along the left side muscle of my chest. No longer can pushed and I stopped, I remained holding the pain and the compressor for a while, when the pain it was a little bit less; I continued pushing and finally I could get the compressor to its place on the platform.

Today, at so many years of that event, I regret it greatly. From that day, when the weather is cold or I exposed myself to a cold place, it gives me a pain like someone put in, a knife in the flank; and, there is nothing that I can do, but to take some pain-killer and to curse, by that error of mine.

At that time, I continued going to teach at night at the National Skills Center school, everything went well, the only problem is that all the students in the classes that I have taught, requested me to return to work with them. The school has had problems with this request from the students.

I went to speak with Mr. Gillian (studies' Director) and I suggested to him, that to avoid problems, he better keep me working with one class only. He accepted and he assigned me to a class that just began; from then on, I just worked only with this class.

===

EXPERIENCE 46-B

<u>Analysis/Procedure</u>

My investigation began in the electrical installation, of the suspected "faulty" compressor.

<u>Step 1.</u> *I disconnected from the lower part of the three-phase contactor, the electrical installation (wires) from the "Faulty" compressor.*

Then, by using my electric voltage meter (proper scale), took the voltage reading from the circuit breaker upper part (inlet); the meter gave the same reading of 440 volts between the "L" terminals 1-2, 2-3 and 1-3.

Brown watches the process doubtful, like thinking that we were wasting the time!

He was standing behind me observing the entire process over my right shoulder.

Then, I pulled up, closed the circuit breaker switch and took the voltage readings in the lower part of the same.

I obtained the following voltage readings:

Between "T" terminals 1-2 = 260 volts. 1-3 = 440 volts, 2-3 = 260 volts.

<u>Conclusion:</u> *"The circuit breaker device was damaged."*

<u>Step 2.</u> *I pulled to the "Off" position, the power supply to the contactor, and manually pushed to the closed position.*

Proceed to verify the contactor: *Set the analog meter in ohms (K1) (Ω -1).*

Place one meter's prong on top (L's terminals), the other on the bottom (T's terminals) of the contactor; the meter showed the following readings:

Between terminals L1-T1, L2-T2 and L3-T3 = The same "0" or "Continuity" reading.

With these readings, all three sets of contacts in the contactor; were OK.

"Meaning the contactor was OK."

Step 3. *Then leaving the ohmmeter set in the same scale (K1) (Ω -1), proceeded to verify the unit motor/compressor's windings:*

Taken readings between the electric motor connecting wires, separately:

1-2, 2-3 and 1-3; obtained the same equal resistance readings of about 2 ohms.

"Showing that the motor's windings were OK."

Step 4. *Then immediately afterward, the ohmmeter was set to the highest OL. M Ω scale (x 100 000 ohms).*

Join together the three wires that fed the unit.

Place the meter's prongs:

One to the three feeding lines together, and the other prong to one pipe of the refrigeration system.

Result: *"No reading." "No ground." "Motor-compressor was electrically OK."*

Brown could not believe what he saw.

Step 5. *I stopped the compressor that was working and in that circuit breaker switch I connected the "defective" compressor.*

I read the FLA reading from the compressor's nameplate (The LRA reading should be somewhere between 3 and 6 times the LRA reading) set the ammeter for these ampere readings, the motor-compressor starts, and works with normal readings.

Still, there was something else left, to be verified. The compressor, mechanical performance. I needed a manifold and some wrenches, I asked Brown and he went out to bring them.

Step 6. By using the appropriate method, I connected the manifold's gauges to the working compressor; and verify its mechanical condition. System's operating pressures.

Here: Big problem! The pressure gauges showed, that although the compressor was running, suction pressure didn't go down.

If this operating condition remains, the compressor does "not pump" normally and it must be necessary to change or repair it. (I assumed that something was preventing (the closing, seating of the compressor's internal valves). Still, I had something left to try.

This procedure may damage the compressor's internal parts, but I had to take the risk. While the compressor was running, by using the service wrench, I closed - front seated the service suction packing gland valve) and let the compressor running for a while (about 10 minutes), then quickly, I opened the suction service valve, and let the suction refrigerant rush suddenly into the compressor.

Then, luckily, whatever was preventing the closing of the internal valves (dirt, foreign matter, whatever) was removed by the risky procedure.

Then the compressor began to perform normally. Brown was so excited up to this point, that he did not notice, this last part of the process.

The only thing that he said to me was "Jose, let me call the company; I would like to know their reaction to this!" I responded to him. "Yes, you can do it, but not here. I do not want this people to find out that they have been by several weeks without the complete air-conditioning service simply by the fault *of a circuit breaker switch. He went out of the premises, made the phone call to the office, and got the authorization to buy a new circuit breaker switch.*

We went to gather the circuit replacement part, to a store located at the Queens borough. Brown asked me, to let him drive me; to buy the device. The cost of the new switch was $29.00. So, we had to buy two devices, because the minimum of the purchase had to be $50.00.

Summary: Time of the repair (including the trip to gather the new replacement device) 2 hours.

If we had changed the compressor, installations of liquid and suction lines' filters, refrigerant's charge, work, etc. (8 hours, two men), the cost would have been, at that time, more than $5,500.00.

==

EXPERIENCE 47-B

The valve located at the condenser's inlet is called "Queen" valve. It is used to isolate the condenser from the compressor in some type of procedures or repairs.

Analysis

This repair implies (one of two (2) procedures):

(1) *To accumulate, by pumping down, the refrigerant content in the condenser-receiver,* or
(2) *To remove the entire refrigerant charge from the system.*

The first option was the most advisable. However, I didn't know how much refrigerant the system holds and it was dangerous to gather the refrigerant in the condenser (I was not sure it has enough capacity). I did not think so.

To empty the liquid line and isolate the part that required the repair; I would pump-down the system by closing the "king" valve at the condenser-receiver's outlet; although, will be risky and somehow dangerous, without knowing if the line was empty of refrigerant but, I must try it. I did not have any other alternative.

Procedure

The pressure gauges were installed, the (king) valve at the condenser-receiver's outlet was front seated, closed; let the system run for several minutes, observing the operating pressures. The suction pressure went down, but not enough and the discharge pressure kept risen, reaching dangerous higher readings; what it was indication that I cannot perform the work.

<u>*I suspected two conditions:*</u>

<u>*One,*</u> *the condenser's outlet valve does not close completely, although I retied strongly.*

<u>*Two,*</u> *the condenser-receiver was not big enough to hold all the system's refrigerant charge.*

I had all the required materials ready. Trying to bring the suction pressure down, I got a partially full R-12 (30 lb) cylinder, from the truck, and try to remove some refrigerant from the system. I got some out, but not much (maybe 8 or 10 pounds) anyhow, the suction pressure descended a little bit more.

Looking for another alternative, I pried the water regulating valve's spring, to let the water go constantly through the condenser-receiver; trying to minimize the pressure on the high- pressure side. This action helped a little bit more. Now, the suction pressure descended a few more pounds.

Having no more alternatives, I decide to take the risk, and I started loosening up the TXV's flare nut, source of the leak. Great problem the line was full of liquid and the liquid refrigerant leaving burned my hands; the worse thing, I couldn't breathe. I tighten the flare nut again; I left the place in a hurry, once outside I took a deep breath of fresh air. I got back and wide opened the place's door and window.

Solution

I disconnected the valve one more time and with some refrigerant still leaking, I was able to remake the flaring, and I reconnected the valve again. Thanks to God, I could finish.

I checked for leaks and got the system back in operation, put back the refrigerant that I had in the cylinder, but, is was not enough, now the problem is that I needed refrigerant; I called to the operator-answering service, and request it to communicate with the other mechanic on service, and ask him to bring me at least three 30-pound cylinders of refrigerant 12. After a while (one and a half hours) I got it. When the refrigerant arrived, I was able to complete the repair.

The only thing that did not please me and I felt very much sorry, was that the cookies that the man was making, they all were damaged (six trays), with the phosgene gas which was formed in the furnace (when burning the leaking refrigerant). My console is at least; he never knew it from where the problem came.

===

EXPERIENCE 48-B

Procedure

I ask him if there were any printings, or plans for that equipment. He said they probably were up there, at the equipment location.

He guided me to the access of heating system location:

This was a gas furnace, at the ceiling above the space to which it served. The gas supply was verified and found OK. Somebody was struggling with the electrical installation, and left "everything disconnected" it was a mess, it was like a big plate of colored spaghetti, at least I found the electrical diagram, and I was able to remake the installation.

Solution

It took me about 2 hours, but I could put it back to work. The man was very thanks full, he offered me lunch and after signing the service order, he gives me a $100.00 tip.

I returned to the city at about 5:00 p.m.

===

EXPERIENCE 50-B

After the initial investigation, I reached this conclusion:

The equipment's heating parts that had been changed did not and have never had anything to do with the place's high temperature.

All A/C systems, besides the regular equipment, controls, cleaning air methods, etc., must be supplied with some sort of fresh air. In this equipment, I saw a metallic air ductwork coming from above a door (on the wall), about twenty feet from the A/C system, entering the equipment through its upper back (a fresh-air duct).

Analysis:

I asked the person in charge if it was possible for me to get to the other side of that door where the fresh air ductwork was coming from. He answered that he must call the building's superintendent.

He went away, and a few minutes later a man (dirty and perspiring) appeared by the door.

As soon as the door was opened, an air even hotter than the air in the room blew in. The space on the other side was the building's boiler room. I explained the problem to the man, then I stood in the doorframe. I looked upward and saw that the fresh air ductwork had been cut, so when the A/C fan worked it sucked in the hot air from the boiler room.

I asked the superintendent if he knew something about it. He told me that at the beginning of March some other company's people had done some work here.

I asked the restaurant's manager if I could use one of the cardboard boxes that they had there. He answered yes. I went to the truck and got a four-inch-wide duct tape roll. I cut the suitable size of cardboard and with the tape I contained the hot air intake above the door.

For a few minutes I put in operation the A/C equipment; the temperature in the place was lowered without any problem. When I asked the individual to sign the work ticket, he was very pleased and even offered me coffee or whatever I wanted. I was not very happy with his initial treatment, so I did not accept anything. It took me about twenty-five minutes to do the "repairs."

I got in touch with the office and let them know my work-findings so they can take the appropriate steps for the proper repair.

As I was leaving the place, one of the company's supervisors entered, satisfied with the temperature in the place, I guess. With a smiling face he asked me if I had finished; I answer him yes. He did not ask or say anything else, and I left.

===

EXPERIENCE 51-B

No need to make any other questions, the mechanic knew that the printings and electrical plans were located at the ceiling on the equipment's covers.

Procedure - Solution

With the aid of the electrical tools (Voltmeter-Ohmmeter), diagrams and the mechanic's help, we were able to pinpointed the problem. We were capable of remaking the installation, finishing it with sufficient time; before the customers start to arrive.

After we finish the job, the owner invited us to have lunch; and he himself served me. He was so happy, that he asked my name; to request the office that from now on, he's word "I do not want to see nobody else but you"

When we were leaving, he gave me a bottle of French wine and he put something into my shirt's chest pocket; to the mechanic he also gave something.

We left. Once in the truck, the mechanic confided to me that the client had given him a $10 bill tip; he asked what he had given to me. I look in the pocket and saw a $100 bill.

Time of work to fix this problem; was about 1 and a half hours.

===

EXPERIENCE 52-B

<u>Diagnosis/Procedure</u>

To verify the complaint, I compared the reading of the wall thermostat with the one registered in my thermometer; I found a very small misalignment. Without the customer's attention, I removed the thermostat's cover and fit it (adjusted the thermostat's needle) to the correct reading.

A little bit later, the customer came back. He reviews it and now he was satisfied. I felt sorry for the mechanics; now they do not have the opportunity of taking advantage of the Chinese man anymore.

===

EXPERIENCE 53-B

Analysis

The first day, the equipment ran well; the second day it ran for an entire hour with no problem. We cannot blame the electrical power supply. So, to me, the high reading of the operating pressures and amperage; may have to do with the condenser's operation.

I went outside to see the condenser, and notice that at this hour of the day; in Amsterdam Ave. the wind blows (uptown) toward the geographic north.

Particularly in this equipment; that, was not good. Vertically, from one-half of the condenser the air was going out (South side) while on the other half (North side) the air was coming in.

In other words: The leaving warmer air from half of the condenser (South side), was returning to the other half of the condenser (North side) creating the high discharge pressure, as well as the high amperage.

Solution

So, we may find the way of separating the two air flows. I thought that a piece of sheet metal (18" x 9") placed in a perpendicular position with the condenser's surface, should separate both air flows currents. We did that and now, the equipment worked without any problem.

Then, I have the problem with the student; he argued that the money of the transaction belonged to him. Thank heavens that the students explained to him, that the equipment no longer belonged to him and; I do not how, but they managed to convince him.

On a later time, the student confided to me, that he had picked-up that A/C equipment from the street. I assumed, the equipment was thrown away because didn't work properly, due to the peculiar condenser air flow "problem."

==

EXPERIENCE 54-B

While waiting for the suction pressure to rise, before conducting any other procedure, I observed the entire system with more detail.

Analysis

More important, when I initially connected the manifold's hose to the small service valve on the compressor's body; I noticed a temperature difference, between the suction line coming into the compressor's flange and the compressor's body itself. This temperature difference gave me the idea, of a pressure drop (refrigerant expansion) through the flange; which indicated a restriction in that place. I also put my hand on the condenser-receiver out-let; there it was no temperature difference there.

I loosen a little the suction line connection flange to the compressor's body, and liquid refrigerant came out, this assured me of that restriction.

Procedure

I proceeded to close the king valve (front seat). After many short starts, the operating pressures remain the same but, I noticed the temperature dropped from the condenser- receiver's king valve to the compressor's inlet flange; which was an indication, of the evaporation of the liquid refrigerant in that portion of the system. I decided to continue with the same procedure.

Due to the almost complete restriction, at the flange; and to help the refrigerant to pass through that point; I used a rubber mallet and hit the flange. It helped but, not much.

This procedure last for about two hours; until I thought there was; no more liquid refrigerant in the suction line for a while.

Solution

At certain point, I by-passed the low-pressure control to allow the suction pressure to reach "0" psig. After getting at this stage; a few more starts were performed. Then the flange was removed completely. Between the flange and the compressor, a filter was found. This device was almost entirely clogged. The filter was removed, and by using carbon tetrachloride, was cleaned and returned it to its place.

The king valve was just cracked open and closed again. This was to increase the pressure in the suction line and be able to check for leaks. No leaks were detected. A partial vacuum procedure was performed on the compressor and the suction line. It was needed not to add refrigerant but to remove it (about 75 lb) because the system had an excess of it.

I assumed the mechanics never installed the discharge pressure gauge for that reason they added refrigerant; moreover as the suction pressure dropped, they assumed the lack of refrigerant and the cycling of the low-pressure control. Time spent on this service was a little over three hours.

==

EXPERIENCE 55-B

<u>Analysis</u>

As this refrigerant was not popular, and I do not have any in my truck; I had to call to the office to request a cylinder of it, as well as a new liquid line filter-dryer.

While the refrigerant arrived, I remove the refrigerant remained from the system, and a vacuum procedure was performed. When the refrigerant arrived, I charged some refrigerant and while it was leaking, I changed the liquid line filter-dryer, charged refrigerant and finish the service call.

Note. *It is ALWAYS very important, that before giving service to any refrigeration or air conditioning system; to make ourselves, 100% sure of the type of refrigerant used by the equipment; as well as getting as familiar as possible with all the electrical and mechanical components of the system.*

Different ways of verifying the refrigerant in an A/C system:

1. *ID plate of the system.*
2. *TXV sticker's color.*
3. *Through P.T. Chart, compare existing refrigerant pressure in the system.*
4. *Etc.*

<u>Calculations:</u> (R-500)

According to the characteristics of this refrigerant, the operating pressures should be as follow:

a) *Discharge pressure:*

Condensing temperature = Ambient temperature =   80°F.
                                          + Temp. diff. (Δt - Dt) = <u>25–35</u>°F.
                                                                 105–115°F.
Discharge pressure = 153–177 psig.

(BP) Boiling temperature = Required temperature =   72°F.
                                        - Temp. diff. (Δt - Dt) = <u>35–30</u>°F.
                                                               37 42°F.

b) *Suction pressure = 41–48 psig.*

The reference point's values should be:

Since the application is the same A/C the temperatures should be as follows:

a) Suction line temperature = Approximately +55 to 65°F (*)

(*) Manufacturers recommend, the SH value to be no less than 20°F at about 6" from the compressor.

b) Liquid line temperature = Approximately 90°F.

==

387

EXPERIENCE 56-B

<u>Analysis/Procedure</u>

I prepared my tool (a multi-meter that I always had with me), make sure the equipment was energized, set the instrument for the voltage value used by the control circuit; in the equipment (24 volts). I inclined myself, removed one by one the covers of each control and the lockout relay, and by touching with the meter's testing prods the two screws in each of them, I found (voltage reading) that the lockout relay's contacts were opened, preventing the equipment from working. All the other controls were closed, thus giving no reading in the electrical voltage instrument.

The result of this test gave me two options:

1. *Make a short circuit between the two terminals of the opened relay's contacts.*
2. *De-energize the equipment.*

<u>*I chose option 1.*</u> *By using a small piece of wire, I shorted-circuit the relay's contacts, and the equipment started.*

The supervisor turned around and asked me what I had done. I explained to him what I had found. The opened device was the lockout relay contacts. (By looking in the equipment's installation/plans, I thought that the opening of a control probably activated this relay. Might be the high-pressure control in response to something wrong in the system's high-pressure side.

Then he said to the mechanic. "You see? He found the problem and it could start after two days here. Why could you not?

The mechanic answered that it was because *he "did not have one of those instruments."* Immediately after about five or six minutes from my arrival, the supervisor said to me, "OK! Jose, you can go home. Thanks."

Option 2. *When using this relay (*) to reset its contacts, the equipment should be de-energized.*

(*) Usually, its sensor a coil is connected in a parallel circuit with the high-pressure control.

Its contacts are connected in a series circuit with this control.

Note: The lockout relay was found defective.

===

EXPERIENCE 57-B

According to the information submitted by the person in charge, he mentioned a job performed by a mechanic "upstairs," so I thought "something related to the condenser."

Procedure

As I already mentioned, the air-cooled condenser was equipped with two fan motors. In my way to the controls box, I put one hand alternatively, on top of the two condenser fan-blades (in order to feel the airflow's direction and intensity), immediately noticed, that one of them was turning in the opposite direction; air flow was downward.

Solution

I stopped the condenser's fans and noticed that the fan with incorrect rotation was a new PSC motor. All I had to do was go to the control panel, find that motor's running capacitor and exchange the electrical connection. See diagram.

In the diagram. the solid line (-----) shows the right connection.

With this, the motor rotates in one direction.

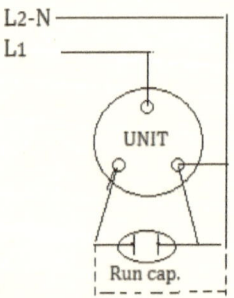

The dotted line (- - - -) shows the wrong connection. With this, the motor rotates in the opposite direction.

This shows me what was the work performed by the first mechanic.

The entire process took about five minutes. I replaced the covers and started again the set's operation.

When I return by the window (remove the shoes), the secretary said to me "Already?" I answer to her Yes.

Then, I went back to the compressor's room. The pressure gauges show, now: a lack of refrigerant. I went to the truck, got an R-22 refrigerant cylinder, add the necessary one; verify the complete operation and I finished the call. Time used here. About 50 minutes.

I do not think I have to mention. But, every time I finished a job, I called the office and let the dispatcher know about my findings and applied solutions. I also noted the work done in the green pages.

==

EXPERIENCE 58-B

Analysis

To repair this refrigerant leak, by its location, it was not an easy one. But we mostly do it. That is what we are here for. It is our work.

Procedure

The system was equipped with a dual pressure control, which I "jumped-out" and by closing the receiver's outlet "king" valve, I pumped down the system to about zero "0" psig. And I prepared myself to make the repair.

Since I noticed that, the insulation on the internal walls of the equipment; was very dry I used an asbestos sheet to separate the insulation from the torch flame; trying to prevent an accident. After preparing the tubing for soldering and applying the correspondent flux, I ignited the torch; I was ready to apply the flame to the point where the leak was located.

Suddenly in a fraction of a second, the insulation produced a flash and was set on fire. Luckily, I had the refrigerant and quickly turn the cylinder upside down; I applied the liquid to the fire. I could control the fire very fast, but not before the place, what I say place? the entire floor got full of smoke. Thanks heaven, that nothing greater happened; it was only necessary to open the windows to remove the smoke. Other than the scare, the only damage was the burned insulation. Otherwise, the system could get back to work without it.

Those people of the radio station commented, that the fire did not propagate to the acetate discs; that if that had happened, it had been disastrous.

Solution

I was able to complete the soldering procedure, then, checked for leaks, pulled vacuum to the low-pressure suction line and compressor, open the receiver service valve again and complete the charge of refrigerant to the system.

The entire system operation was verified.

===

EXPERIENCE 59-B

<u>Analysis</u>

Well, this may happen. Sometimes, I guess, to save some money, some people risk their reputation and acquire the wrong material in the wrong place. Our experience tells us that tubing designed for R & A/C applications has, among others, a special characteristic: "No joints" "No porosity" and to get this special feature, manufacturers make it (melted); with a very fine grain.

To prevent any type of refrigerant leaks, this tubing is put under very special treatments. Once it goes to the market, it comes in especial packages and in their containers the manufacturer says all the especial characteristics of it.

Cooper is cooper. Yes. BUT the R & A/C copper tubing is different, to the ones manufactured for carrying other fluids. I.e. air, water, steam, oils, etc.

I call the office and reported the problem; they were going to speak with the radio station management; to discuss the repair.

===

EXPERIENCE 60-B

Analysis

The problem was studied and reached the conclusion, that as this machine works with R-502 when the compressor stops; the low-pressure rises approximately to 210 pounds (98°F. or 36.7°C.). This compressor did not have protection against that so high pressure, and the problem took place. The so high pressure did blow out the crankcase's sight glass.

Calculations (R-502)

According to the characteristics of this refrigerant, the actual operating pressures are the following:

a) *Discharge-pressure:*

Condensing temperature = Ambient temperature =   98°F.
+ Temp. diff. (Δt - Dt) = <u>25–35</u>°F.
123–133°F.
Discharge pressure = 291–331 psig.

b) *Suction-pressure:*

(B. P.) Boiling temperature = Required temperature = -10°F.
- Temp. diff. (Δt - Dt) = <u>-15</u>°F.
-25°F.

The normal B. P. for -10 °F. (Δt. /Dt. = 15°) should be:

Desired temperature   = - 10 °F.
                            - 15 °F. = - <u>15</u> °F.
               B. P. for R-502 = - 25 °F.   = 57.5 psig.

Suction pressure = 57.5 psig

With a conversation to the customer, it was explained the problem and recommended to him the installation of a water-cooled condenser as well as a "Suction" or "Crank-case Pressure Regulator" (SPR or CPR).

He agreed with the change and gave the OK for the repair.

Solution:

To order and install:

1. A 1/2 of a ton capacity, water-cooled condenser equipped with a (WRV) Water-regulating valve.
2. New sight-glass replacement.
3. Install a new SPR Suction Pressure Regulator (or CPR, crankcase pressure regulator).
4. A liquid line filter-dryer replacement.

The recommended water-cooled condenser was of a coil-within-a-coil type; it was equipped with a water regulating valve. The new water-cooled condenser was installed on top of the existing air-cooled one; it was connected in a series circuit with the air-cooled condenser.

The new crankcase sight glass and the pressure regulator mechanical component were installed as well.

The threaded liquid line filter-dryer type was replaced with tubing with the same dimensions.

The entire job was verified for leaks. A vacuum procedure was performed for about 45 minutes. By using this vacuum effect, the oil charge was accomplished. By adding R-502 the vacuum was broken, the pressure raised to up to 5 psig. While the refrigerant was leaking, a new liquid line filter-dryer was connected. The refrigerant charge continued, until the operating pressure was normal and the equipment worked properly.

Water lines to/from the condenser were installed; the WRV was adjusted to start opening, when the discharge operating pressures was, the maximum allowed with the air-cooled condenser.

Now, the discharge operating pressure should be at a point, between the Air and Water cooling mediums temperatures, used here.

Air = 98°F. Water = 60°F. Combined = About 79°F.

New Discharge pressure:

Condensing temperature = Air/water temperature =   79°F.
+ Temp. diff. (Δt - Dt) = 25–35°F.
104–114°F.

Discharge pressure = 227 to 260 psig.

The entire system's operation was verified.

Note. Now, with the addition of the new water cooled condenser, the amount of refrigerant used in this system was a little bit larger; to compensate for the new added component.

===

EXPERIENCE 61-B

<u>Analysis</u>

To me, this guy's behavior did not seem to be honest at all; but, I did not say anything to anyone, because; I thought that should be my word against his and I, who was new and did not dominate the language could put myself into a problem.

Sometime later, in my school; I face myself with the same problem, this time with a former student who confides to me the same situation, and to whom this time I was able to reprimand and objected this poor and dishonest form to behave.

===

EXPERIENCE 62-B

<u>Analysis/Procedure</u>

Since this system used water as a cooling medium, and as I mention before, the water supply as well as the return were "warmer" and it was a little temperature difference between them . . . I bent myself to check the condenser's water connections, and note the problem:

The condenser's water connections were wrong; they should be connected "Counter-flow"; this means (Water in - refrigerant out). In my opinion, this problem was not new; it had been there for several years; since the installation of the old equipment.

Despite the condenser had marked (IN - OUT) and arrows indicating the water directions flow. Whoever perform the first installation, fail to accomplish the "Water - refrigerant counter flow"; that it is supposed to be, in order to get the appropriate heat exchange between the two substances (Cooling medium - Refrigerant) for the system to operate properly.

This time, the same mistake was repeated.

Changing the system or component parts did not solve the problem!

Without saying absolutely anything to the customer, I left the pharmacy to direct myself to where the cooling tower was located. When I got there, note that everything was in order, rotation and operation of the water pump, fan motor, water system cleanliness, etc. Only problem, water was a bit warmer due, I thought, to the water temperature control adjustment.

From the place where the cooling tower was located, till the place where the two (2) water (supply/return) lines went up to the A/C system, it was a distance of about fifteen (15) or twenty (20) feet.

I study the way to "fix" the problem without the customer realized it. The only possible solution: To invert the water connections. Took the piping's dimensions; Diameter, elbows, pipe extensions, nipples, in general everything that I believed was needed to perform the job.

Since I did left the company, so far it has been an hour. Call the office to speak to the supervisor and to communicate him my "findings." He could not believe it. As I never was that very good plumber, I requested for someone to come and complete that part of the job. He told me that he would send a plumber immediately.

I went to a nearby hardware store, bought all the needed materials to make the change, and took them to the tower location.

I went back to the Pharmacy and told the customer that we had to do something in the cooling tower. He said to me. It is OK.

About half an hour after my call, the plumber appeared. I went with him to the cooling tower's location and explained to him, plain and simply what had to be done. No mystery. Then add. "Prompt, please, do the job as fast as you can."

Again, I went one more time to the pharmacy, des energize the system, I amused "wasting" time, giving the opportunity to the plumber to complete the work before the client noticed. About half an hour later, I went back to the cooling tower.

<u>MY MISTAKE</u>. I did not check the work. The plumber told me that it ended in five minutes, that I could already put the system to work.

Back at the pharmacy, I put the equipment to work, but, when I touched "felt" the pipes' temperatures, I noted that they showed the same temperatures, than when I started my investigation. I stopped the system again.

Returned to the cooling tower location; I noticed that the "plumber" had only changed the piping sections, never inverted them. I did not want to fight with him (after all I was the one in charge), so I had to make the change by myself.

When the inversion was almost finished, a supervisor appeared. He wondered why the delay. He thought that I had already ended. Endure the terrible timing of the supervisor and I said only; that I had had a small setback but already could energize the equipment.

With the supervisor, we got back to the Pharmacy to put the system to work.

Now, the system started and it worked properly. In front of the supervisor, the owner, delighted, congratulated me. I had to get back to the cooling tower location; to verify the aquastat control adjustment (it was a little higher), it was adjusted.

Case concluded. The entire job, including the mistakes, took about two (2) hours.

==

EXPERIENCE 63-B

Analysis

I suspected that something was wrong in the place's electrical installation. I recommended, reviewing the wiring installation at the power outlet, where the refrigerator was connected; for it to work properly (the wire of that point must be of not less than # 12 AWG) but, if he wanted to try the refrigerator again. Then, before connecting it, I suggested verifying with the voltmeter the voltage value in the outlet; in addition, to maintain the voltmeter connected, and when the refrigerator was putted ON; to observe the voltage value. This value does not have to descend more than the 10% of the voltage value reading, taken before putting the refrigerator to work.

According to the equipment's ID plate (F.L.A. 3.2), the L.R.A. in this device should be about between 9.6 and 19.2 amps.

Procedure

I also suggested; if the refrigerator started, let it work for at least 10 minutes. Then, disconnect it or shut it off from the thermostat; and after about 2 or 3 minutes, put it back to work. Always observing the voltage value reading. I asked him to call and let me know, what happened.

Later on, he called me. He said; he venture himself and without watching the installation wire size, he took the refrigerator and did what I had suggested; the voltage reading in the place was of 110 volts. When the refrigerator was connected the first time, the voltage descended to a value of 95 volts. The LRA registered was noted, took a bit long but the refrigerator started and kept working.

After 10 minutes of working, he stopped the refrigerator and 3 minutes later, it energized again; but, this time the voltage value descended to 80 volts; what was more than the allowed 10% voltage drop. The refrigerator did' not started and began to cycle again, by the electrical thermal overload.

Explanation.

Now, why this problem happens? In the initial starting, the equipment's operating pressures were equal, although the voltage descended more than the allowed 10%, the motor-compressor did not experience greater problems, and it could start.

After 10 minutes operation, the equipment created the operating pressures; when stopped (2 or 3 minutes, time more or less normal to start again by the thermostat), the operating pressures do not have enough time to equalized; the motor-compressor set, cannot "overcome the operating discharge pressure opposition" and fails to start.

The voltage drop is reflected by the wire size. I.e. wire # 14 is good for about 15 amps.

Trying a household refrigerator with these characteristics, to work in that location; should imply, to change the electrical installation in the entire building. I told Anthony to suggest the client to get a 1/8 hp refrigerator instead of the 1/3 hp. The customer refused the idea.

Here, I think, it is not necessary to say that the business was undone.

===

EXPERIENCE 64-B

<u>Analysis</u>

Many people try to repair the auto-air conditioners, by using the already known motto: "Do it yourself" buying refrigerant in Strauss and other automobiles parts warehouses; without telling that aside from the "charge" of refrigerant, there are also; other reasons for the badly or the non-operation of the equipment.

In this endeavor we even had two cars of the year which (according to the customers) the dealer's people had them in their places for several days and had not found the problems. But we were able to find them (refrigerant leaks), and we sent the clients to buy the needed spare parts (gaskets) in a warehouse of the branch.

We also review-repaired small equipment; window units (with these systems we work with R-22). Final result everybody happy and satisfied. We collected, near 50 pounds of refrigerant excess (in those days R-12), between all the repairs that we carry out. My wife did the times of treasurer and at the end of the day, it report a good profit, which we share with my collaborators (students).

This, it was the work of a single and only one day. I did not want to get myself into trouble, with the neighboring auto shops. Anyway I did not waste my time.

<u>Calculations</u>: (R-134a)

According to the characteristics of this refrigerant, the operating pressures should be the following: Approximately.

a) *Discharge pressure:*

Condensing temperature = Ambient temperature = +/-85°F.
+ Temp. diff. (Δt - Dt) = <u>25–35°F.</u>
+/-110–120°F.

Discharge pressure = 148 to 171 psig.

b) *Suction pressure:*

(B. P.) Boiling temperature = Desired temperature = 72°F.
- Temp. diff. (Δt - Dt) = <u>30–35°F.</u>
42–37°F.

Suction pressure = between 32 to 37 psig.

(R-22)

a) *Discharge pressure:*

Condensing temperature = Ambient temperature = +/- 85°F.
+ Temp. diff. (Δt - Dt) = <u>25–35°F.</u>
+/-110–120°F.

Discharge pressure = 226 to 260 psig.

b) *Suction pressure:*

(B. P.) Boiling temperature = Desired temperature = 72°F.
- Temp. diff. (Δt - Dt) = <u>30–35°F.</u>
42–37°F.

Suction pressure = between 63 and 72 psig.

(R-12)

Note: No need for calculations for this refrigerant, since it is no longer in use.

==

EXPERIENCE 65-B

It was about 4:00 p.m. when I arrived at the store; located at Nereid and White Plaines avenues in the Bronx. I found that the ice cream machine was not working, and all the ice cream had been damaged (more or less 150 bars of chocolate, vanilla, etc.).

The operating characteristics of this machine had nothing to do with the ones of the systems with air-cooled condensers located in the basement. This was a self-contained unit with an air cooled condenser, located at the store main quarters.

My client, had his "man of confidence," that had not been able to identify the problem (a lack of power); by using a light bulb. He asked me for help. By using my voltmeter, within a few seconds, the problem was identified; a faulty circuit breaker.

Now, back to my work.

Analysis

From my first visit to this place, I had recommended the use of three (3) supplemental water-cooled condensers; for the three air-cooled condensing units working in the basement.

This day, I had the three (3) water-cooled condensers (Coil-within-a-coil type) to be installed.

It took me about 2 hours, for the refrigeration work, which included: Recovering the refrigerant from the three (3) systems, one by one:

Since all the systems were in the same place, and all of them used the same refrigerant R-12 and belonged to the same owner; by using a five hoses (*) manifold, I removed the refrigerant, on the first machine; and put it in the other two systems. When the pressure was 0 psig. in the first system, I located the water type condenser, on top of the air cooled condenser and connected in a series circuit; with the air cooled condenser below. Checked for leaks, pulled a vacuum and returned the refrigerant, until operating pressure were normal and the equipment worked properly.

(*) Myself adapted this manifold; I added two extra valves. Now I had:

1. One ¼" hose to be connected to the high-pressure side of the system.
2. One ¼" hose to be connected to the low-pressure side of the system.
3. One 3/8" hose to be connected to the vacuum pump.
4. Two 1/4" hoses to be connected to: a refrigerant cylinder, or oil container, or recovery machine, or other system, or Etc.

I repeated, exactly the same procedure with every one of the other two systems; until all three were working properly.

The water supply installation was performed, the three supplemental water-cooled condensers, were equipped with their corresponding water regulating valves.

The water valves were adjusted to start opening, when the discharge operating pressures were, the maximum allowed with the air-cooled condensers.

Calculations. R-12

1. *Discharge pressure.*

   The operating pressures were calculated, in all three systems, taking into account the combined two cooling medium temperatures:

   a) Ambient temperature   = 90°F.
   b) City water temperature = 60°F.
                         150°F. divided by 2 = 75°F.

Condensing temperature = Combined temperature =   75°F.
+ Temp. diff. (Δt - Dt) = <u>25–35</u>°F.
100–110°F.

a) Discharge pressure = 117–136 psig.

2. <u>*Suction pressure.*</u>

These operating pressures were calculated, according with every one of the requirements in each application.

Example: Beer cooler.

(B. P.) Boiling temperature = Required temperature =   40°F.
- Temp. diff. (Δt - Dt) = <u>20–15</u>°F.
20 - 25°F.

b) Suction pressure = 21–24 psig.

Besides the original refrigerant charge, some refrigerant was added to all three systems, to compensate for the installation of the new water-cooled condenser. Then, the entire three systems operation was verified, they were left working perfectly.

===

EXPERIENCE 66-B

<u>Analysis</u>:

The noted operating temperatures (suction and liquid lines), the pressures (discharge and suction), and the cooling tower water supply differential were all higher than normal. It was indication of an abnormal system's operation.

The customer complaint was clear:

1. "The system very often takes too long to start cooling."
2. When it finally does, he thinks the cooling effect is not "as good as expected."

    a. It started cooling in the morning, at the beginning of the day.
    b. But it was not cooling enough for the rest of the day.

Why? Now, the system is cooling although it wasn't normal.

For me, the problem started with installation. So let's begin by getting familiar with the system and finding out "what was wrong."

After verifying separately every component of the A/C system, I reached the conclusion:

1. The entire electrical installation's components had nothing to do with these problems.
2. The refrigeration system operating pressures and temperature conditions were higher than normal.

Other conditions noted: Space temperature a little too high for this hour.

All these conditions, together, were indication that something was wrong.

Now, the only next operating condition to be verified was the cooling medium.

So, let's check the water system, as well as find out an explanation for the spilled water in the basement.

When verifying at the condenser's supply/return water lines temperature, it was noted that the water regulating valve was connected to the condenser's water outlet. This line was located at the upper part of the condenser.

In the basement, it was noticed as well that:

1. The cooling tower was located below the equipment's condenser, a distance of about 15–20 feet.
2. The water-circulating pump wasn't equipped at its outlet with the required check valve.

Diagnostic

With the water flow direction, created by the wrong WRV installation, and the lack of the check valve, when at the end of the day the equipment was de-energized, the condenser was completely drained out. The cooling tower overflowed, creating the spilled-water problem.

At startup time in the morning, the water-empty condenser may have created a high pressure, which activated the high-pressure control, probably a couple of times, until the condenser was full of water, and the system started to work and to cool.

This wrong WRV connection failed to comply with the counter-flow refrigerant vapor-liquid and water-flow direction required to accomplish the equipment right operation, and explained:

1. The abnormal operating conditions: pressures, temperatures
2. The sight-glass bubbles, and
3. The poor cooling effect.

Solutions

1. Install a check valve at the water pump's outlet.
2. Change the water valve location to the condenser's water line inlet.
3. Adjust the water-regulating valve.
4. The entire refrigeration system operation was verified, and some refrigerant was added.

After obtaining and installing the required components, the equipment's entire operation was verified.

==

EXPERIENCE 67-B

Analysis

Then, I suggested measuring the voltage at the device's power supply, when the same were energized. Thus, he did it and according to; the voltmeter gave a reading of "only 55 volts."

I told him that, it looks; he had connected the device in a series circuit with another electrical component, probably the electrical heater in the room. He verified the electrical connection, and noticed that indeed, he had connected it in a series with the named device.

Use of electrical meters {Voltmeter, Ammeter, and (Analog Ohmmeter)} it MUST be a VERY important custom. With their readings, we may have the real electrical information, and with that, the best guide to performing any electrical job.

===

EXPERIENCE 68-B

<u>Analysis</u>

Usually, cooling towers are devices cleaners of air. The cooling tower's water curtain, removes from the air passing through, any particulate matter flowing with it. This effect is used in some A/C systems of certain bigger capacities. Although the use of this cleaning method is limited, because increases the moisture content of the air.

Yes. Sometimes, filters can be used to that cleaning purpose; although, the fine matter manages to get thru and that's the one, more difficult to remove from a piping system.

Suggestion

I elaborated a small diagram and I spoke to the engineer; I ask him if he ever hear about a "settlement tank"? He said no. Then I suggested to him, the use of a chamber or separating tank, which may be located at the cooling tower's water outlet; to trap "catch" the undesired element before entered the water-circulating pump, then; the water system.

This settlement tank may be equipped with draining valves, and it could be easily and frequently drained and cleaned out as necessary. The idea was accepted.

===

EXPERIENCE 69-B

Therefore, I decided to make the calculations on my own.

<u>Calculations</u>

<u>*Meat characteristics:*</u>

1. Specific heat above freezing point = 0.77 Btu/lb.
2. Freezing point   = 28°F.
3. Latent heat value at 28°F   = 100 Btu/lb.
4. Specific heat below freezing point = 0.41 Btu/lb.

<u>Calculations:</u>

Sensible heat, latent heat, and total heat.

A. Qs. = W (Δt.- Dt.) Sh.
    5,000 (40 – 28 = 12 °F.) (meat = 0.77 Btu's/lb)
            5,000 x 12 x 0.77 =   *46,200 Btu's.*

B. Ql. = W (LHF)
    5,000 x 100   = *500,000 Btu's.*

C. Qs. = W (Δt.- Dt.) Sh. (meat = 0.41 Btu's/lb)
    5,000 (28 - -15 = 43 °F)
       (meat = 0.41 Btu's/lb)   = *88,150 Btu's.*

· 15 % = Leakage through walls, doors, people, etc.

Total = Qs. = 46,200
       Ql. = 500,000
       Qs, = 88,150
          = 634,350 + (15 % = 95,153) = 729,502 Btu = 40,528 Btu /hr.
                     18 hrs. 12,000 Btu/hr./ton
                         = +/- *4 Tons.*

°F

— 40
A
28 _____ 28
B
28 _____ 28
C
_____ 0
-15 _____ -15

System characteristics

1. A semi hermetic compressor 4-ton capacity with air-cooled condensing unit.
2. Electrical requirements 230 volts, 1 phase, 60 Hz.
3. Due to the required low temperature the system used R-502 (*).
4. Dual type pressure control

(*) Note. *Today, this system may use the refrigerant replacement for R-502.*

With the refrigerant's characteristics and exactly the same calculations.

Operating pressures

Calculations (R-502):

According to the characteristics of this refrigerant, the operating pressures should be as follows:

a) *Discharge pressure*

> Condensing temperature = Ambient temperature = +/-85°F.
> + Temp. diff. (Δt - Dt) = 25–35°F.
> +/-110–120°F.

Discharge pressure = 246.3 to 251.5 psig.

b) *Suction pressure:*

> *The temperature differential (Dt - Δt) for low-temperature application could be equal to 15°F.*

Desired temperature = –15°F.
Temp. diff. (Δt - Dt) = –<u>15</u>°F.
                    = -30°F.

BP for R-502   = at -30°F   = 9 psig.

The price agreement was reached and the equipment was installed.

The job was completed and the equipment work perfectly.

==

EXPERIENCE 70-B

<u>Analysis</u>

According to the customer's, Robert's information, system characteristics as well as the showing symptoms; it seemed, this problem; was caused by a liquid refrigerant return. I asked, for any spare removed parts. Robert said the only part was the valve that he had removed. The customer said he hadn't seen any other parts around.

When Robert showed me, the expansion valve that he had changed/removed; I saw it and noticed that this valve was of the internal equalizer type; it seemed to me, that for a system of this characteristics; the expansion valve should be of the external equalizer type.

As it was not possible for me, to get to the TXV and Evaporator's location; I asked Robert to go and to look if at the evaporator's outlet there was a connection. After checking it, he came back, and said that; at the evaporator's outlet; it was a 1/4" capped connection.

I asked him to go to the warehouse, where he usually bought the spare parts and buy a same size characteristics valve but, with and external equalizer.

Solution

During the pumped-down procedure, it was necessary to remove some refrigerant because the discharge operating pressures was too high. When the proper pressure reading (0 psig.) was obtained, the TXV was changed as well as the liquid line filter-dryer; the system was checked for leaks, a partial vacuum process to the evaporator-suction line-compressor was performed.

When the system was putted back into operation, its operating pressures were a bit too high, therefore; some more refrigerant was removed (all together, about 50 lb). The pressure controls were readjusted.

Finally, the system's operation was verified and the equipment work without any problems.

Explanation.

The problem arose, when the thermostatic expansion valve was mistakenly changed.

As we may remember, these valves are operated by three (3) forces:

1. Opening = (1) Thermal bulb pressure.
2. Closing = (2) Sh. Spring pressure, plus
                (3) Evaporator pressure (external equalizer).

With this mistaken valve, the pressure (3) was absent. So, the valve flooded the evaporator creating the "frozen" condition and subsequent problems.

Operating pressures.

Calculations: (R-502)

According to the characteristics of this refrigerant, the operating pressures should be as follows:

a) *Discharge pressure:*

Condensing temperature = Ambient temperature = +/-85°F.
                         + Temp. diff. (Δt - Dt) = <u>25–35°F.</u>
                                      +/-110–120°F.

Discharge pressure = 246.3 to 251.5 psig.

b) *Suction pressure:*

The temperature differential (Dt - Δt) for low-temperature application, it could be equal to 15°F.

(B. P.) Boiling temperature = Desired temperature = -20°F.
- Temp. diff. (Δt - Dt) = <u>15°F.</u>
-35°F.

Suction pressure = 9.0 psig.

==

EXPERIENCE 71-B

Analysis

According to the symptoms, the problem could be due to the opening of, usually, the high- pressure (*) control; apparently both were of the automatic reset type. I thought that beside these controls, the system might have another control: A lockout relay.

With this last control, when a problem appears, and it triggers the control to stop the system it affects the relay as well (by opening its contacts); now, the equipment would not starts again unless is completely de-energized. When this happens, the relay its reset and the equipment can work again.

I suggested to him to use, a "thread" of a fine wire connect it in a parallel circuit with the high-pressure control's electrical connection. This would take him not more than a minute. After this, called the company and tell them that he finished but ask them to let him return on the following day.

In addition, I suggested to him that if on the following day he found the wire in the control "melted" (an indication that the control had opened), he should concentrate all his attention on finding what it caused the discharge pressure to raise.

When —— called me back, I said. "Look in the printing. Find the high pressure control and see if there is anything else (coil, resistor) connected in a parallel circuit with it."

He said it was a coil as well as some contacts with it.

I asked him to shut off the equipment and using the ohmmeter to take the reading on the device's contacts.

He did it. And the meter gave a resistance reading.

I told him that device had to be changed. It was damaged.

After two days, he called me; he was happy and he told to me that the problem it had been created by air flow through the condenser due to the fan motor, slow (lubrication), cut-out setting of high pressure control too low (220 psig.), as well as some accumulation of dirt on the condenser's surface.

(*)   *Personally, I do not like this type of control.*

A lockout relay may be used with the same purpose as the manual reset high-pressure control.

As cited, when a too "high" discharge pressure condition appears due to:

1. *Airflow (**)*
2. *High-pressure control' setting etc. (***)*

The condition doesn't disappear by itself. It requires attention of a technician.

The right high-pressure control should be of the "manual" reset type. My opinion.

(**) *This type of problem (aggravated by the temperature at the time of day, may have other reasons that could alter or affect the airflow through the condenser:*

1. *Accumulation of dirt on the condenser's surface.*
2. *Fan motor - Loose electrical connections, or lack of lubrication.*
3. *Hot air from another source. Check if, lately, any new system had been installed-located near this equipment's condenser.*
4. *Verify the adjustment {Hi event (Cut-out point)} of the high-pressure control.*
5. *Etc.*

(***) The high-pressure control should be set, by using a temperature differential (TD. − Δt.) and the formula-calculation shown in Experiences 05, 17 or 60 (The R-22 PT numbers) can be used.

With an ambient temperature of about 85°F. a higher discharge pressure should be of about 260 psig. I usually applied a pressure differential of +/-25 psig. This would give a cutout (Hi event) point on the high-pressure control of about +/-285 psig.

===

EXPERIENCE 72 B

The comments for this experience, I think are self-explanatory. The important things to stand out, was the way the different jobs were performed. As well as the learning that can be obtained, from the varied performances, carried out.

The behavior of the mechanic wasn't new for me; it was nothing else to do, but to cut the job as early as I could. It wasn't the first person here, acting with me like that. So . . .

==

EXPERIENCE 73-B

<u>Analysis</u>

I suggested, other than verifying the power supply; to check the running capacitor and I remembered him, about the tool and testing procedure to be performed. Minutes later, he called me back to communicate me, that the problem was a "short circuited" running capacitor.

The faulty device was changed, the installation of the new unit as well as the rest of usual procedures were performed. Then, the equipment worked without any more problems.

Again, electrical instruments. Without them, it is almost impossible to properly diagnose an electrical component/equipment.

Usually, these capacitors are of the low MFD range; 2, 3. 5, 7 1/2 15, 20, etc. The MFD's value depends on the system's capacity.

To test them two methods are used:

PRECAUTION: "Electrical circuit MUST BE de-energized."

(1) A capacitor's tester (Use Manufacturers information) and
(2) An ohmmeter with a scale x 1 K (1,000) Ω (See Information below)

NOTE: *Make sure that paint, rust, dirt, etc., do not interfere with the meter's leads connections. (Clean or remove as necessary)*

Run capacitor testing. Measures:

(1) MFD values.   (2) Openness.   (3) Shorted.   (4) "Ground."

Analog ohmmeter.

Note: *For me, the analog type of ohmmeter is the best and most accurate electrical instrument, to be used in the R & A/C fields, to take or verify resistance - ground conditions in any electrical device.*

Procedure

1. Adjust the meter to Zero ("0" Ohms) reading.

After preparing the specific meter and selected the appropriate range:

a) Put the meter's leads together.

b) Verify, that meter's needle moves all the way to the other end of the dial.

c) By turning the "ADJUST" knob on the meter, align the needle directly, over the "0" mark on the meter's dial scale.

NOTE: *If the needle does not reach the "0" mark on the dial, the batteries may have to be changed.*

d) Repeat step "c" each time, before using the meter or whenever change meters rangers.

(1) MFD values, *this meter does not give a specific capacitance value. When the capacitor is in good condition, accordingly with its MFD range value, the meter's needle must deflect toward the "0" mark (Higher the MFD value, the higher the needle's deflection) then, all the way back, to the "infinite" (*∞*) initial position.*

(2) Openness. *In a capacitor with this condition; the meter's needle dos NOT move.*
(3) Shorted. *When a capacitor has this problem, the meter's needle should deflect and remains static near, or on the "0" mark of the meter.*
(4) "Ground." *This condition is tested between the capacitor's terminals and the metal container. Any needle deflection may be an indication that, the capacitor plates are touching the device's enclosure or have some type of a defect.*

===

EXPERIENCE 74-B

Comment

The only !Problem! the mechanic had; was that he used too much Teflon tape, to secure/ connect the valve to the condenser's water lines connections. And due to this, when twisted-connected the valve, only few threads were involved in the work of securing the valve to the piping.

Next day. Seven o'clock in the morning: Terrible news. Disaster; during the night and due, not only to the work defect; but, also to the pressure and the water weight; the water valve's connection got loose from the piping; and as a result the place was flooded; damaging all the equipment and archives that they had in the computer's room.

At about 8:00 a.m. at the regular hour of start the day's work, the mechanic came to it. But, before he got into the company's Office, he found out, from another mechanic, about the problem. He told him wherever had happened and, as the refrain said, "He took those of Villa Diego." In other words, he disappeared. For me, he went with his "knowledge" to some place else.

Several days later, I hear out that the computer's company owner, had filled a lawsuit against ——— by 100 million dollars!

In the following months, even the WRV's manufacturer company was involved, in the problem and litigation. From then, they limited the guarantee of its product.

I did not know the rest of the history, neither the end of the problem.

The whole Experience is more or less self-explanatory. Some people pretend to "know" when in reality, they do not have the minimum requirements to show it.

I only hope, that someone; may learn something from this experience.

==

EXPERIENCE 75-B

Analysis

For the high pressure, obtained through the manifold reading and the opening (tripped out) of the high pressure control, he blamed the water-regulating valve (WRV) and decided to change it.

After using the time, in "troubleshooting" and obtaining and changing the water valve; approximately three and a half hours (3 1/2) had passed, and when he was almost finishing the valve's change, the establishment's owner arrived and, asked him what he was doing?

He answered to him! Changing this water-regulating valve! He asked him why? The mechanic answered: "Because it does not allowed the passage of water through the system!

And, as a result, the equipment wasn't working.

Then the owner said to him, that in the place there was no water, because the utility company was performing a repair in the street.

==

EXPERIENCE 76-B

<u>Analysis</u>

I asked him what the cut-in adjustment was, and what temperature was at the two condensers location.

He answered, that he had adjusted them at the same cut-in point of 60 psig. (R-22). Beside, the actual place's temperature was 30°F.

I suggested to him, to take a look at the P/T chart and to fit the controls cut-in points for 50 or 52 psig. that it corresponded to about 26 or 28°F.

He made the adjustments for both equipment and after that, they both worked well.

Note. *Whenever the condensing unit of any refrigeration or air conditioning system, is located in a place where the ambient temperature is low, it is necessary to have that fact into consideration.*

Due to the existent low temperature, the pressure affecting the Low-pressure control operation should be low and the system may be off. That condition requires setting the cut-in point at a lower but safety pressure value, than the regular set point.

===

EXPERIENCE 77-B

Comment

Another one and we have so many of them.

YES. *We will always have something to learn. It does no matter how old we are or how much we think we know or from whom but there will always be something unknown by us, that it can be learned to enrich our intellect, and be placed in our vast brain.*

Usually, every A/C or R system has in its ID. Plate, information about the required testing pressure for each system's side. I.e. R-22 High pressure side 300 psig. Low pressure side 150 psig.

Every time the system is placed under pressure testing, it's greatly advisable to be sure about the gas being used for such test.

===

EXPERIENCE 78-B

Comment

In the school, through my students, I remembered and hear from various persons, having problems in different occasions; with some compressors; when they went to get an ex-change they had a hard time, with the person at the store's counter. But, when they used their knowledge, and told him the date when the device was manufactured, then the compressor change was given to them, without further problems. At least, they get the exchange with no problems.

The service technicians should learn to read (interpret) the information given in the (ID) plates of each product. This way, the least he can know (among other things), is the date of manufacture of the product.

It is very convenient to have in our shop or working place, a compressor's testing stand equipped with pressure gauges and etc. So, when the time arrives and we have to verify the working condition of a unit (Compressor-Motor), it may be necessary to start it and to let it run; to check its amperage (Start and Run) and the suction and discharge conditions; before the installation at its working place. This test, if performed; should be a fast one, to minimize, as much as possible, the set's contamination.

==

EXPERIENCE 79-B

<u>Analysis</u>

When any mechanical device is designed, the manufacturer calculates its connecting tubing or piping's size; according to the amount of substance to be required, to pass through them.

In the field, we can observe and use the size of that tubing or piping joining connection; as a reference to attach whatever needed to make or complete our job. Any smaller size should create on the fluid's flow, a pressure drop; while an oversized implies, usually, among other things, more work and wasting of money since bigger the tubing/piping, the higher the cost.

Example

One (1) ton refrigeration - air conditioning system, requires certain specific amount of water in the condenser, to remove the heat load from the refrigerant.

Let's assume that the system (hermetic – semi hermetic) of our calculation works in an environment where the ambient temperature is 80°F. Uses R-22.

Condensation Temperature = Approximately 85°F.

As we may know, the amount of heat to be removed (condenser) from the refrigerant is equal to about:

Heat from evaporator   = 12,000 Btu/ton/hr.
Heat from Super-heat (approx.) + = xxx Btu
Heat from the Electric motor   + = 2,545 Btu/hp (*)
                Total heat (approx.) = 16,000 Btu/ton/hr.

$$\frac{16,000}{60 \text{ min. /hr.}} = 267 \text{ Btu/ton/min.}$$

(*) Electric motor used in this case 1 hp (746 watts) (1 watt = 3.412 Btu)
                                    = 746 w. x 3.4 Btu/w. = 2,545 Btu

By using the appropriate formulas/calculations, we can find out, how much water is needed to accomplish the normal heat removal; through the condenser job in this A/C 5 tons unit.

"FORMULAS and CALCULATIONS"

$Qs. = W (\Delta t)\ Sh.$

Where:

Qs = Sensible heat (Btu/lb) W = Weight (pounds)
V = Volume (cu ft) Δt = Temperature difference (°F.)
Sh = Specific heat (Btu/lb/°F)

From the formula: $Qs = W (\Delta t)\ Sh.$

$$W = \frac{Q}{(\Delta t)\ Sh} = \frac{267\ Btu/ton/min.}{(12)\ 1\quad 7.4\ lb/gal} = \underline{22.25}\ lb/ton/min\quad \frac{22.25}{7.4} = approximately\ 3\ gal/ton/min$$

When using water from a cooling tower, it is customary to use a temperature differential (Δt) of 10 to 12°F.

Discharge pressure:

Condensing temperature = Ambient temperature (*)   =   85°F.
                         + Temp. diff. (Δt - Dt) = <u>10–12</u>°F.
                                                          95–97°F.

Discharge pressure = 182–189 psig.

(*) *Some adjustments may be performed to combine the temperatures of air and water.*

==

EXPERIENCE 80-B

Since all the resistance readings from both windings were normal, it occurred to me to start the unit manually, so I did it with the help of the starting capacitor. The unit started and worked perfectly. Both LRA and FLA amperages values were more or less according to the compressor's ID plate and the machine operation.

Since the potential (voltage) relay was not able to disconnect the starting capacitor, the thermal overload stops kicked the unit out. This made me concentrate my attention on the motor's operating induced voltage.

Analysis

Once the system was running, with the voltmeter (proper scale) I measure the "induced" voltage reading (between the motor's Common and Start terminals) starting winding; and I compared with the printed "pick-up voltage" value in the relay's cover; for some reason, the induced voltage reading, was lower than the one required by the relay to open the contacts and de-energize the starting winding.

When verifying the capacity of the relay and the unit, they were exactly, one to the other.

Explanation.

I believe, the reason for this "low" induced voltage from the starting winding, was probably due to a small "shorted circuit" in any of the motor's Run or Start windings. This way, fewer wire turns were exposed to create the right "pick up" voltage value.

At starting time, because the induced voltage was lower than the one needed by the relay to open the contacts; this device fails to disconnect the starting winding, keeping the high LRA value too long. Being that the reason for the opening of the thermal overload.

Solution: *To use a relay of a lower capacity (lower "pickup" voltage value).*

We went to the store relay's supplier, and through the manufacturer's catalog; we compare and choose a relay with the required characteristics.

With the new recommended relay installed, the system worked perfectly.

==

EXPERIENCE 81-B

<u>Analysis</u>

By verifying the low-pressure control (cut-in) adjustment, encounter that it was set for 30 psig. That was the answer-reason; why the machine did not work properly.

Taking into account that temperature of the place was very low, we decided to change the "cut-in" adjustment to 27 psig (+/- 28°F). When doing it, the machine it started. This way, the low-pressure control adjusting operating settings were changed, making them normal for the present cellar temperature conditions. Case concluded.

As the reader has noted, through the description of these Experiences, I pursue to give some information, which can be used as a reference. The times have changed as well as some of the refrigerants, oils, component parts, and some tools; but the following remain the same:

1. The refrigeration system operation,
2. Ambient conditions,
3. Products storage requirements,
4. Must of the procedures and many of the tools are still used,
5. And most important of all, the operating calculations are the same applied then, and actually.

So in each of these experiences, a lesson can be learned and it is up to the reader how to use it. Some of these experiences are very simple and very easy to apply, while others are a little bit more complicated, and so they require a little more of study, analysis, and attention.

==

EXPERIENCE 82-B

Comment

Water temperature and other technical characteristics of these types of equipment; can be easily obtained, through one of the various warehouse suppliers of equipment and implements of these systems.

As a reference, the required amount of refrigerant used here, should be the same as any other 3/4 of a ton capacity system; using R-12. Experiences 02 and 10 can be used as a reference for the operating pressure conditions and the amount of refrigerant.

Explanation

In this experience, as in some others; I think is not necessary to make any additional comments, since the whole procedure was explained in detail. In the case of a system like this, the equipment's manufacturer informs the amount of refrigerant charge (by weight) so when the system is put into operation the appropriate operating conditions appear and the system produces the particular required conditions.

===

EXPERIENCE 83-B

Analysis

According to the compressor's temperature, the system had being cycling or working intermittently; besides, the difference in temperature across the SPR device and the actual pressure reading shows something wrong:

1. The installation of the SPR device or
2. The location of the L.P. control

Diagnosis:

The SPR device had to be removed. It was not necessary here. Usually, this SPR device is used when a system has a semi hermetic compressor. Its purpose is to protect the compressor against excessive operating suction pressures. It was never of popular use when the system has R-12.

Solution:

SPR was removed, the system checked for leaks, some refrigerant was added, and systems operation was verified. The equipment was back to normal operation.

===

EXPERIENCE 84-B

Analysis

To complete the test, the voltage reading between the two supply lines (above the fuse) was taken, and the reading was equal to 220 volts. I went to the first floor and asked the owner since when was this compressor working with this installation. He answered me, "I do not know, probably for a couple of years!"

The new compressor with the proper information was obtained.

Solution

After disconnecting the burned-out compressor, the entire system was submitted to the following procedures:

1. By using R-11 and pressurized nitrogen, the equipment was decontaminated.
2. The filter-drier was replaced with two flare nuts and a piece of copper tubing the same size.
3. The same procedure as in number 2 was performed in the suction line at about one foot away from the compressor for a new suction line filter-dryer installation.
4. The new semi hermetic compressor was installed.
5. After, a careful check for leaks was performed.
6. The system was submitted to a two-hour vacuum procedure.

The system's vacuum was broken with R-12 when the pressure was positive, about three or four pounds, and while the refrigerant was leaking, the new liquid and suction lines' filter-driers were installed.

Charging the system with the proper amount of refrigerant until obtaining the normal operating conditions completed the service.

Other Comments

No logical or technical explanation for this compressor's operation with as much as twice the voltage value supplied to it. Just incredible!

==

EXPERIENCE 85-B

<u>Analysis</u>

When the refrigerator's door was opened, there were no sounds (whistles) indicating the peculiar passing of refrigerant through the liquid control (gurgling) or boiling in the evaporator.

Note: *These sounds are always present, during the normal-entire operation of any R & A/C system, although in many systems, these sounds are not audible because other noises interfere with them.*

<u>Procedure</u>

One plate on the freezer side was removed to allow the access to the location where the capillary tube entered the evaporator. Then, by using a match, its flame's heat was applied to this point; immediately, it felt like the ice deposited at the end of the cap's tube melted, and a sound (whistle) was heard of the refrigerant passing through, then gurgling, boiling, and it started producing the cooling effect (frost) on the evaporator's surface.

Note: *This problem is NEVER presented in A/C applications. Why?*

<u>*Two reasons:*</u>

1. *The normal operating refrigerant's BP is never below the freezing point of the moisture in the air. Dew point.*
2. *R-22 has more affinity in mixing with moisture than other refrigerants, and the effect is less probable with it.*

Diagnosis

"Moisture in the system." After about four years of normal operation.

Conclusion

The drying agent (desiccant) lost its effect of "retaining" the moisture and it loosened it, letting it circulate together with the refrigerant through the system.

I thought nothing lasts forever! So, with the drier-agent here.

Note: *As cited above, I experienced the same problem twice in my service life (See Experience 10). Within a span of about ten years from one event to the other. The first one was overseas, and this one here in the United States.*

Coincidence? I do not think so. The refrigeration system was and is the same.

==

EXPERIENCE 86-B

History Development

The day before, I planned to call and order the coil's construction; for some reason that I do not remember, I had to go to a scrap R & A/C equipment "warehouse" located in the so-called El Barrio in Manhattan in one of the 100th Streets between 2nd and 3rd Avenues.

In this place, I saw a lot of old equipment. The business of these people was to negotiate with scrap metals. Out of inquisitiveness, I asked them if they had a copper coil with the dimensions that I had; they told me that they were going to see, and they would call me if they found something.

Two days had passed by, and I received the call. They said they had found one copper coil that might comply with what I was looking for.

I went to the warehouse immediately; the coil they had was practically new, and according to my calculations, it could fulfill the demanded operating characteristics. The old coil, it was only a little less wide (approximately four inches) than the one that I needed. But everything else—there was no problem; it was only necessary to attach to it the connecting heads from the old coil.

I spoke with the person in charge of the scrap place; he suggested mounting it and supplementing the lacking space with a flat piece of metal that would cover the opening, which I did. They supplied me with the needed piece of metal and also offered to buy the old coil. The transaction was completed, and the coil was taken to the school. The connecting heads from the old coil were removed then attached to the new one.

After the installation of the heads was performed, a water leaks check was carried out. Two manual valves were joined to the coil's piping for draining purposes. The job was finished, and the customer was satisfied.

==

EXPERIENCE 87-B

On these types of A/C systems, the easiest and fastest way to verify the electrical conditions is to submit the equipment's electrical components to an evaluation through the electrical resistance readings taken from the equipment's connecting plug.

Procedure (*Resistance Test*)

Remember! The higher the used voltage, the higher the resistance values readings.

Note

This testing procedure is very simple. No part of the system had to be disassembled. And not much work is involved, although it requires the technician to be aware of the equipment's electrical configuration.

For this test, the information (legend) enclosed within the diagram (next page) was used.

By using an analog ohmmeter instrument, scale "X1" Ohm (Ω -1), the meter was adjusted to zero reading, then the following procedure was performed:

1. The ohmmeter leads can be connected to the "L1" and "N" terminals at the equipment connecting prong's plug or to the selector's switch inlet terminals.

    Note: *It was made sure the selector switch was in the OFF position and the thermostat switch settled on the ON position (closed).*

2. The ohmmeter gave an "∞" infinite (no resistance) reading.

3. The selector switch was moved to fan "Low" speed position, the meter read = _17 Ω_

4. The selector switch was moved to fan "High" speed position, the meter read = _12 Ω_

5. The selector switch was moved to "Cool," fan "Low" speed position, the meter read = _8.2 Ω_

6. The selector switch was moved to "Cool," fan "High" speed position, the meter read = _6.9 Ω_

The sealed motor was disconnected and tested, and the ohmmeter gave the following readings:

1. Between "C" and "R" = ____1 Ω__

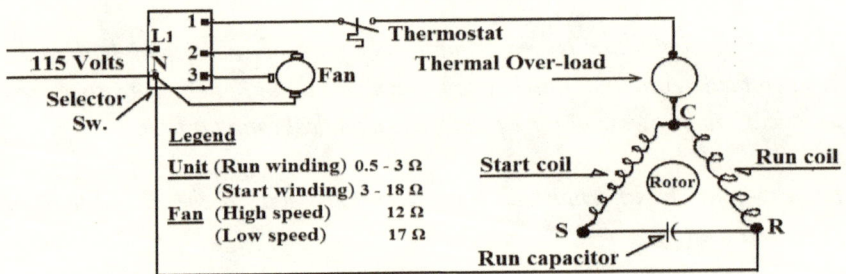

2. Between "C" and "S" = 16 Ω.

3. Between "R" and "S" = 17 Ω.

According to the actual resistance values, the readings on numbers 5 and 6 (Page 1) were wrong, which indicated the running capacitor (from factory) was wrongly connected. That was why the compressor did not work.

In other words, the connection "N" in the diagram above, instead of being connected to the unit's "R" terminal (as shown), was connected to the unit's "S" terminal.

Note: *If the resistance readings were taken at the electric plug before all this procedure, we may notice the wrong high readings in numbers 5 and 6. In any R or A/C system (single phase), those readings should be low, probably, between the resistances values of 0.5 and 3 Ω.*

Taken into account the unit and fan motor resistance values registered inside the diagram, the following values should be the right ones:

The normal resistance readings (seal unit and fan motor) at number 5 must be = 0.94 Ω. and the reading at number 6 must be = 0.92 Ω.

Reference

The electrical installation in these equipment shows a normally parallel circuit formed by the running electric motor's winding and one of the fan motor's speeds (high-medium-low).

The electric law says:

"The total resistance value in a parallel circuit MUST be lower than the value of the lower resistance's component in the circuit."

==

Notice: *The starting compressor/motor's winding is NEVER present in the resistance readings from the cord plug because the connected running capacitor is a natural "opening" in that circuit.*

==

EXPERIENCE 88-B

I always thought that it is better to perform every job without the customer present. Not to be dishonest, but I prefer not to be interrupted in my concentration and work.

Analysis

By using the appropriate tools, I found out the following:

1. Motor's electrical conditions: ohmmeter and voltmeter

   a) The motor was not burned-out or grounded.
   b) Both LRA and FLA amperages were normal.

2. Compressor's mechanical condition: manifold's gauges

   a) The compressor "did not pump."
   b) It had to be repaired or changed.

With the result of this analysis, I went down to the director's office to inform him of what I found. I indicated the problem to the director, and he authorized me to find out the cost of the replacement or the repair.

Thus, I did it; the following day, I called and let him know the cost of the work:

Two alternatives:

(1) A new unit cost was about $3,000.
(2) The cost of a repaired one was $1,500.

In addition, the cost of my work was $800.00.

It was necessary, to contract a crane to first lower the damaged compressor and later to raise the repaired one. The total cost of the crane was $300.

Of the two options, the second (the most reasonable) had a total cost of $2,600. He told me that he would think about it and would call to let me know his decision.

A little more than a week passed, after which he called me back and told me that he accepted the price with a repaired compressor. The following morning, with the help of one of my students and using the appropriate tools, the "king" valve at the receiver tank's outlet was closed. We tried to perform a "pumped-down" procedure.

We isolated the compressor from the system by closing "front-seated" both compressors' service valves. After the compressor was removed, its connections were covered with plastic bags. Once the compressor was craned down, using a cart, we carried it out to the school that was only a few blocks away.

When we arrived at the school, I decided to take a risk. I called one of the wholesalers of the compressor's manufacturer company, and by using the information from the machine's ID plate, I asked for the spare parts set's price. The cost of the complete valves plate set, as well as the entire gaskets set, was $225. I ordered the parts, and I went to gather them to the Bronx. I brought them, and together with the students, we opened the compressor and we found that the internal valves (suction/discharge) were, as I suspected from observation of the thermostat, broken.

Since I did not want to expose any parts of the unit to moisture or the ambient conditions, we started the job immediately. The broken valve's pieces, as well as the entire old gaskets' set, were properly and entirely removed. The locations of all of them were completely cleaned, and the new sets were installed.

After less than three hours, the job was finalized; by using two plates that I had in the school, we covered the compressor's suction discharge connections. These plates had means to connect to them the manifold gauges. Taking all types of precautions, the unit was energized (440 volts - 3 phases) only for few seconds, and the operation (pumping effect as well as the amperage values of the unit) was tested. It worked perfectly. After this test, the oil charge was completely drained out and the amount measured.

After the repair and test, the compressor mechanical plate's connections were sealed (capped) to prevent the ambient air from coming in.

The compressor body was properly cleaned and painted. I let the paint dry, and after two days, the compressor was returned to its working place. The crane came back to raise it, and it was installed.

Then:

a) A test for refrigerant leaks was performed to the entire system.
b) A deep vacuum of about two hours was performed to the isolated compressor.
c) By using the unit vacuum, the proper oil charge was added.
d) The king valve was cracked opened, and while R-22 was leaking, the filter-dryer at the liquid line was changed to a new one.
e) Some more refrigerant charge was added.

The entire perfect system operation was accomplished.

The client was advised and explained of the thermostat's low setting.

===

EXPERIENCE 89-B

Comment

After the accident, it was verified that the cylinder's content was not nitrogen, but oxygen.

"Oxygen will explode in contact with oil or grease."

Whenever we work with compressed gases, we must exercise the highest care of not mixing them with other ones, as is the case when working in refrigeration or air-conditioning systems.

A lack of care when testing, mixing, charging gases, etc., may cause very dangerous situations, not only for the operator, but also to others, as well as the equipment in which we work.

We must always remember the compatibility of the substances involved in a mixing process.

The testing pressures to be used depend on the type of refrigerant used in the system. This information is usually found on the ID plate of the system/compressor.

==

EXPERIENCE 90-B

<u>Analysis</u>

By using the valuable information from the customer, I performed a fast but meticulous inspection of the system. Since every mechanic that verified the equipment did not find any refrigerant leaks in the usual places, i.e., flare nuts, connecting parts, etc., and this system uses water as a cooling medium, I thought in this system, the problem may be in one of two places: the condenser or the pressure relief valve.

The first and easier one was the pressure relief valve.

When these valves open to relieve an excessive pressure, there is almost always a trace of oil left in its opening port. This can be "detected" visually or by finger's touch. This trace of oil usually leaves some amount of refrigerant, and when the leak detector is used, the device is able to register that trace. Sometimes, these valves do not even return to a complete close. Here, there were no oil traces, and when the leak detector was used, no leak was registered.

This condenser (shell and tube) type had the two water connections in one of its ends; it was equipped with shut-off valves in each line, as well as draining valves. The condenser's design meant water passed twice through it. The water piping connections were attached to the condenser by a flange and screws. The other condenser's end was sealed.

Because the system was in operation, by using the proper procedure to verify the operating pressures, the service gauges were connected. Both pressures were low, which may indicate a lack of refrigerant.

Some refrigerant was added to keep the system working and to have enough pressure (discharge side) to detect the refrigerant leak.

The drain valve in the condenser's water exit side was opened; some of the leaving water was collected in a small bucket. A leak detector (Halide-torch type inspection hose*) was passed over the water's surface; it showed (what I suspected: the change of the blue color of the flame (normal) to the green color (leak), showing the refrigerant presence in the water sample. It indicated that it was in the condenser's tubes, where the refrigerant's leak was located.

*A leak detector of the electronic type could be used with the same leaking refrigerant results.

A report of the findings was made to the client, as well as to the company's dispatcher.

Due to the type (design) of the condenser, it had to be changed. According to the dispatcher, a company's supervisor will get in touch with the customer to make the repair arrangements.

===

EXPERIENCE 91-B

Comment

For me, this form of work was not only a waste of time, electrical energy, and most important of all, polluting the system since through that leak and the created vacuum, not only air (foreign matter with it), but also moisture was introduced to the equipment.

On one occasion, and without anybody noticing it, I went into the machine's room where it was one of the compressors with this problem, and I began to look carefully after some indication that showed something to me.

After a short while, I noticed a small pool of oil on the floor underneath the compressor.

I concentrated my attention and noticed that from the (semi hermetic*) compressor's body, by the point where the electric motor is separated from the cylinders, there was a small hair like line where the oil was coming from.*

When I was leaving, a mechanic entered. I ask to him as how they dealt with the problem. He looked at me in a strange form, like saying, "So? What do you care about it!" Without saying anything else, I lead him to the compressor, and I showed first the oil pool on the floor, then the leak on the compressor's body. He did not say anything to me. And I just left.

Now, my explanation for the first paragraph's use of the phrase "rare form of work." I mean, a lack of coordination or maybe a lack of communication—I didn't know how to name this.

Another example: A group of workers from one company showed up to install the carpets of a supposed future office. Two or three days after they finished, personnel from a different company showed up. They were a group of painters, and they did their work in the place without caring about the carpets. Some other time later, another group—this time, electricians—came to install lights, switches etc., and to accomplish their work, they made holes and such in the walls.

For me, it does not make any sense.

==

EXPERIENCE 92-B

Comment

The damage to the compressor's body was so serious that no repair was possible. In our report, we recommended changing the compressor.

Note

A packing gland-type discharge service valve should NEVER be front-seated or shut-closed while the compressor is running. The most common way of preventing this is by making sure the high-pressure control installation and settings are the appropriate and safety ones; this control should be connected at all times and should NEVER be "jumped-out" as it had happened here.

Of course, there are certain special occasions in which this valve (discharge) has to be front-seated:

1. Not completely front-seated, but very carefully in a restricted position to verify or adjust the "cutout" set point on the high-pressure control.

   Although for this endeavor, we may use some other controlled means, such as restricting the condenser's water/air flow, etc.

2. Almost the same condition as in number 1. When the compressor is verified for its pumping or efficiency conditions.

3. With the system de-energized; front-seat completely to remove the open- or semi hermetic-type compressors.

4. Same position as in number 3 when the compressor has to be isolated in procedures such as:

   a) Refrigerant recovery
   b) Charging or removing oil*
   c) Repairs in the compressor itself, i.e., change or repair of internal valves, plates, gaskets, etc.
   d) Changing or adding parts on the system's discharge side (oil separators, mufflers, etc.)
   e) Etc.

*Although this procedure has nothing to do with the discharge valve position

==

EXPERIENCE 93-B

<u>Analysis</u>

By my experience (In New York City), when a system uses the city as a source of water (temperature approximately +/-60°F), the normal temperature difference (Δt - DT water in - out) is about 20 to 25°F (-7 to 12°C.), a difference that can be noticed by touching alternatively the condenser's supply and return water lines.

I approached, every one of the four systems using this type of cooling medium, and by using the hand-touching procedure, right away I noticed that all of them were showing a temperature difference much lower than the reference noted above.

These conditions indicated the water-cooled condensers were dirty (which interfered with the heat exchange between water/vapor-liquid refrigerant), which increased the equipment's discharge operating pressure, as well as the water flow through the device.

Pressure gauges were installed in one of the systems, and I noticed the discharge pressure was too high. = 130 psig. (Normal should be about 100 psig. [See calculations below])

Due to these factors and higher operating conditions, the water regulating valves (WRV), activated by the high pressure, are opened to its maximum, creating the wasting of water.

These particular conditions imply the need for the condensers to be cleaned.

<u>Calculations</u>

1. Discharge pressure:

   Condensing temperature = Water supply temperature = +/-60°F.
                             + Temp. diff. (Δt - Dt) = <u>20–25</u>°F.
                                                   +/-85–90°F.

   Discharge pressure = 92 to 100 psig.

2. Amount of water

$$W = \frac{Q}{(\Delta t - Dt)\, Sh} = \frac{200 \text{ Btu/T/min.}}{20 \times 1} = \frac{\underline{+/-10 \text{ lb}}}{7.8 \text{ lb /gal.}} \text{/T/min} = \text{Approx. 1.2 gall/T/min}$$

I let the customer know my findings. I also told him that I was going to get some acid compound to clean the condensers. With his authorization, I got the chemical substance, and I proceeded to stop and clean, one by one, the condensers needing that procedure.

At the end of the cleaning procedures, all four systems' operations, as well the adjustment of the WR valves, was verified.

==

EXPERIENCE 94-B

<u>Procedure</u>

After the unit's contactor mechanism was disconnected with the aid of the ohmmeter, we changed the two burned fuses and energized the equipment. We found that the evaporator fan motor was in good operating condition.

At the condensing unit's location, by means of electrical instruments, it was re verified; the compressor's motor burned out as well, as the grounded condition, I suspected, was because of the woodchip in the contactor mechanism. A further electrical verification showed the condenser's fan motor burned out as well.

The condenser's surface was found dirty, with a lot of dirt accumulated on it. I guessed that these two conditions (the condenser's fan motor being defective and dirty) were the reasons for the high-pressure control trip-out.

The dispatcher was informed of the findings, and the replacement parts were ordered.

===

EXPERIENCE 95-B

<u>Analysis</u>

I had to ask the man in charge, "What was the required temperature?"

He answered, "36°F. (2°C.)

<u>Calculations</u>: (R-12)

According to the characteristics of this refrigerant, the operating pressures should be as follows:

a) *<u>Discharge pressure:</u>*

        Condensing temperature = Ambient temperature = +/-78°F.
                    + Temp. diff. ($\Delta t - Dt$) = <u>25–35</u>°F.
                                   +/-103–113°F.
        Discharge pressure = 122.5 to 142 psig.

b) *<u>Suction pressure:</u>*

        (B. P.) Boiling temperature = Desired temperature = 36°F.
                      - Temp. diff. ($\Delta t - Dt$) = <u>15</u>°F.
                                  21°F.
        Suction pressure = 22 psig.

The liquid line temperature should be about 88°F.

Procedure

I proceeded to connect the pressure gauges to both pressure sides through its readings to be sure what the problem was.

Both pressure readings were lower than the required. The sight glass liquid indicator was observed as well, and I noticed bubbles in it. These indicated a lack of refrigerant.

I proceeded to locate the refrigerant's leak. It was very easy; I found it by looking at a small trace of oil on the floor right under the receiver outlet piping connection.

The oil came from the flare nut of the receiver's connection outlet. The flare nut was retightened, but the leak continued.

The system was pumped down until zero lb, and the flaring was remade.

The repaired portion of the system was checked out for leaks.

I pulled an auto vacuum* to the part of the system exposed for the service repair.

*Front-seated the discharge packing gland valve, and the manifold's right-hand valve was open. The compressor was started to pull a vacuum procedure from the king valve location through the evaporator, suction line, and compressor. Air and non-condensable gases came out through the right-hand and center manifold's hoses.

The needed amount of refrigerant was added. The entire operation of the system was verified and the service call was finished.

With regard to this service call, days later, I commented about it with another mechanic. He told me of another mechanic who had done a similar job. He gave me his name.

That service call, was to the amphitheater of Cornell University on Second Avenue. There, they had a walking box where they kept corpses (hanging off by the ears) to be used by the medical students.

He did not have as much luck as I had because in his case, the refrigerant leak was in one of the expansion valves flare nut connections within the cooled space. Of course, he had to get into the space to make the repair. After he finished it, he went home, and it did not showed up for work until after three days.

===

EXPERIENCE 96-B

Analysis

Here, it is important to remember an electrical rule for the parallel circuit:

"The total resistance value in a parallel circuit must be lower,
than the lower resistor's value in the circuit."

Diagnosis

By using the A/C equipment, and following the electrical diagram, and the resistance readings within. Fan coils: {(35, 28 and 24 Ω) and compressor's motor running winding (2 Ω)}

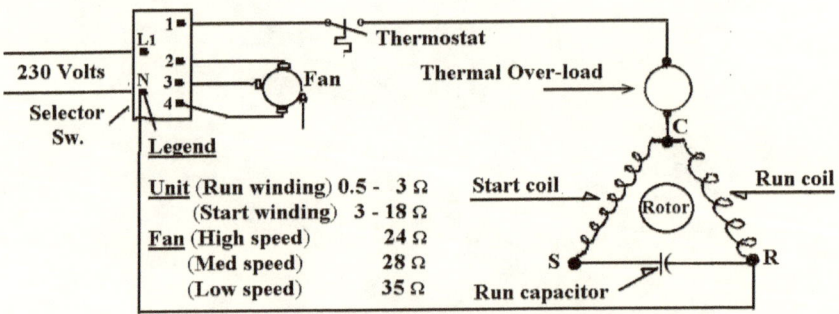

According to the obtained resistance readings, this was the reason why the equipment, although new, was thrown to waste.

The normal resistance readings (unit's motor and fan) should be:

Example:

("d"): accordance with the mathematical formula:

$$\frac{R1 \times R2}{R1 + R2} = \frac{35 \times 2}{35 + 2} = \frac{70}{37} \text{ "L"} = 1.89\,\Omega$$

by using the same formula with numbers
for "M" "e" = 1.86 Ω and "H" "f" = 1.85 Ω

Note: *This makes true the formula described above.*
     The total resistance value on this circuit must be less than two (2 Ω).

So resistance readings at "d," "e" and "f," procedure "3" (first page), were higher, showing a wrong installation, which indicated that the running capacitor (from the manufacturer) was wrongly connected.

At this point, we reconnected the unit, the running capacitor's connection was changed as shown in the diagram, the equipment was energized, and it worked perfectly.

Once more, in any R or A/C system (simple or single phase), the resistance value readings should be low, probably between the resistance values of 0.5 and 3 ohms.

===

EXPERIENCE 97-B

Analysis

Based on other experiences (see Experience 14), I do not recommend repairing this equipment because the system had been badly contaminated. In addition, attempting to decontaminate it would be very difficult; it could end up having the same problem: the burning of the compressor.

During this experience, I intended to call the attention to the R & A/C service technicians to be very careful when it comes to replacing electrical devices (motors, transformers, etc.) in relation to their operative characteristics (especially Hertz), which were manufactured in other parts of the world since for some time, somehow, these devices are available in the United States market.

Through the following information and calculations, I will explain the difference between the frequencies of 60 and 50 cycles (Hertz) in a two poles motor.

Hertz (cycles) 60 - 50 Formulas: rpm (revolutions per minute)

$$\text{Rpm (Sync.*)} = 120 \text{ (Cy.)} \frac{\text{Frequency (Hz.)}}{\text{\# of Poles}} = 120 \frac{60}{2} = 3{,}600$$

$$\text{Rpm (Sync.)} = 100 \text{ (Cy.)} \frac{\text{Frequency (Hz.)}}{\text{\# of Poles}} = 100 \frac{50}{2} = 2{,}500$$

*Synchronous speed

Explanation:

As can be seen here, there is a marked difference in speed between these two electric motors. Of course, the refrigeration cooling effect is obtained "faster" at a higher speed, but the "abuse," will produce a much higher working amperage as well as temperature. And sooner or later, the result is a burned-out motor. While in operation, the moving parts wear away faster, the entire system is getting contaminated, and in the end, such effect is so big that the motor is burned-out and it does not pay to be repaired.

That's the result of wrong Hertz.

===

EXPERIENCE 98-B

Comment

Almost every refrigeration and air-conditioning system has to be prepared, for summer as well as for winter. Included within that preparation is (even in summer) verifying the refrigerant charge.

Solution

Then I recommended verifying the relief valve visually and with a leak detector; he may find the answer for the lack of refrigerant there. The valve may have opened, letting some refrigerant go in order to release the excessive discharge operating pressure. Usually, with the refrigerant coming out, some oil goes out as well, leaving a trace on the valve's outlet port.

This may happen at any time. Sometimes, during the cold weather time (autumn, winter, and spring), the operating discharge pressure in any refrigeration or air-conditioning system gets too low for the system's normal work.

Some air-cooled systems may have their own equipment to control this condition: capacity controls, control valves, condenser's fan motors speeds control, fan motor's cycling, air flow controls (dampers, etc.) etc. But other times, and in order to increase the discharge pressure and to keep the system working properly, the technician had to use certain procedures, such as blocking the air flow and or adding some refrigerant, etc. Then when the warmer weather arrives, the "remedy" has to be removed; otherwise, the adopted procedure does not permit the system to work appropriately. Here, the excess refrigerant must be removed.

If the excess of refrigerant it is NOT removed (by the service person), then the relief valve, if any in the system, does its job: it releases the cause of the malfunction.

===

EXPERIENCE 99-B

Too many questions Right? And only one answer Let's see it.

<u>Analysis</u>

Based from an experience that happened a long time ago (1962) in a beverage company's industrial refrigeration application:

By using the thermostat, I de-energized the solenoid valve and stop the system. I proceeded to disassemble the valve, remove the coil, and noticed:

    a)   The valve's coil was loose on the valve's stem

    b)   Besides, the coil was too short (by about 3/4 inch)

For me, these two problems created the high temperature and the excessive vibration.

What happened here was because of the air gap due to the wrong size of the coil, the space around and on top of the valve's stem was not filled completely. (The coil did not produce enough magnetic field to move the stem to the proper position). This also created the excessive vibration. The current flow as well as temperature increased and the coil burned.

From the valve's ID plate, the printed information was taken, given to the engineer, and we recommended calling the valve's manufacturer to ask for the proper coil size.

From his office, the engineer called the valve's manufacturer; they exchanged information about the valve, and according to what I heard, they said the valve's coil size was wrong. They gave him the coil's replacement data, which he wrote on a piece of paper and gave to me. I told him that my company will order the new coil, and they will send someone to complete the job. He agreed.

I called the company, asked for the supervisor, and render detailed information of my findings, as well as the coil's data.

===

EXPERIENCE 100-B

No comments for this experience

==

EXPERIENCE 101-B

<u>Analysis</u>

The following collected facts were considered:

1. The superheat reading in the medium temperature room (68°F.) was kind of high
2. The suction line temperature (when both rooms are working normally) should be around 40°F. It was now about 60°F.
3. Liquid line at the medium temperature should be about 85°F. It was about 75°F ambient temperature.
4. The temperature at the medium temperature room's outlet was warm. Ambient temperature.

Diagnosis

All of the above operating factors were an indication of a faulty medium temperature room expansion valve operation.

This could be caused by one of the following reasons:

1. *Valve stuck closed*
2. *Valve restricted (clogged filter)*
3. *TXV thermal bulb lost the charge, etc. . . .*

Calculations

Medium temperature room's pressure:

(B. P.) Boiling temperature = Desired temperature = 48°F.
- Temp. diff. (Δt - Dt) = <u>10</u>°F.
38°F.

E P Regulator pressure setting = Approx. 31 psig.

Solution

The thermostatic expansion valve in the medium temperature room was changed. Then, the entire system's operation was verified.

===

EXPERIENCE 102-B

<u>Analysis</u>:

Few but serious considerations

According to the characteristics of these two substances, this change is not possible:

1. In "open" type cooling towers, there is always evaporation. In order to replace the evaporated glycol, it should be not only more expensive but also difficult to maintain its appropriate density.

2. Due to the difference in their specific heat values:

    a) Water = 1 Btu/°F/lb.   b) Glycol 0.85 Btu/°F/lb.

(1) It should be necessary to circulate more glycol than water.

(2) In order to obtain number 1, it should be necessary to change not only the circulating pump size but the entire piping system as well.

<u>Calculations</u>

For calculations, experience 93, as well as the glycol's specific heat value, can be used.

Solution

The only and most effective solution was to return to the original water system.

==

EXPERIENCE 103-B

Analysis

With the electrical circuit energized and using the voltmeter because of the environment (low temperature) and control setting, we found that the sensor was open (115 volts reading), and for that reason, the solenoid valve, timer, compressor, and evaporator fan motor and condenser fan motor were not working.

Solution

Short-circuited (jumped out) the sensor

The solenoid valve, timer and condenser fan motor were energized and the A/C system started.

The system operation was completely verified. Everything was normal.

The director was advised about the taken solution (jumped control) and that in a couple of days (when the temperature increased), we should come back to remove the jumped connection on the ambient temperature sensor control.

==

EXPERIENCE 104-B

Diagnosis

We explained to the client that the aluminum foil leaves were the reason for the bad cooling of the stored product. The aluminum leaves covering the grilles, though it may prevent the metal from oxidizing, obstructs the free air pass flow, preventing the product's suitable cooling.

This was the reason for the low operating pressures and temperatures as well.

Solution

To remove the aluminum foil leaves

Finally, the entire system operation was verified, and we found it normal.

Note: *A long time ago, this same condition had been observed in a domestic refrigeration service, with the same explanation from the client and the same solution.*

==

EXPERIENCE 105-B

Procedure

By using the thermostat settled on the OFF position, the system was stopped.

We pulled out the condensing unit (we noticed the condenser's fan motor was new) and installed the pressure gauges. The system was started again, and the problem was verified: "wrong air flow" and both operating pressures were higher.

Diagnosis

The fan blade was moving the air: first through the compressor, then the preheated air was delivered to the condenser. This wrong operation makes the operating pressures higher than normal and, consequently, causes the system's lack of the proper condenser cooling.

Note: *Normally, the fresh ambient air must go first through the condenser. (This is to accomplish the required, refrigerant - cooling medium counter-flow.)*

Explanation: *Usually, the air at 80°F. goes through the condenser first, then it goes to the compressor to help in its cooling.*

Here, the air went first through the compressor, picked up heat from it, and the preheated air (80°F +) was delivered to the condenser.

As a result, both of the equipment's operating pressures were higher.

The wrong airflow caused the malfunction and the poor cooling of the stored beverages.

The electric fan motors used in the evaporator, as well as the ones used in the condenser, are marked: clockwise (CW) or counterclockwise (CCW).

Solution

(1) *To get, an electric fan motor with rotation opposite to what it was here, or*
(2) *Find a set of blades with opposite pitch than the ones used here*

The second option was adopted; fan blades were obtained and then changed.

The entire system operation was verified, and some refrigerant was added.

On the last services, to keep the discharge pressure "lower," some refrigerant had been removed.

===

EXPERIENCE 106-B

Procedure

After more than about twenty minutes of the harvesting time, I de-energized/ stopped the equipment and proceeded to investigate.

We tried to get to the bottom of the evaporator (where the ice cubes were formed) and, with our fingers, "felt" a large number of the loose ice cubes still there. I felt in this ice mold, as well as in many others, the ice mold's edges were a little smaller than the rest of the ice molds, making the harvest process difficult to perform and, of course, the cubes to drop.

Diagnosis

My opinion: Disfigurement of evaporator's material (cube molds) due to water/ice expansion (at changing physical state) were increased because operating controls set were too low, and the system's cycle operation produced a very low suction operating pressure.

Service time duration: a little more than one and a half hours. Called the dispatcher and gave him the report. According to him, the supervisor had to talk to the customer, who will decide whether to change the evaporator or to buy new equipment.

To verify the operating pressures in these types of machines, experience 40-B could be reviewed.

Note: *Years later, I noticed the manufacturers changed the shape of those molds. Instead of the former square, they were made slanted or a little wider at the edges.*

===

EXPERIENCE 107-B

Diagnosis

As mentioned in part "A," the fan blades were located at the top of the cooling tower, and the air must be moved vertically upward in the opposite direction to the (falling) water movement.
(Air-Water, counter-flow)

Procedure

Since everything looks right, the next step should be to put the equipment in operation. Right?

Well. When the fan assembly was set ON, the problem was noticed:

Wrong fan motor assembly rotation

Solution

The equipment was de-energized.

In the power distribution box, located at the side of the cooling tower; the reverse or exchange of two of the three power lines to the fan motor were performed.

Then the new rotation was verified, as well as the entire A/C system operation.

Some refrigerant had to be added.

==

EXPERIENCE 108-B

Something here does not make sense. There was only one solenoid valve controlling the water flow to both tanks. They both were equipped with floats.

Diagnosis

Here is the reason for the operational problems.

Since no water was coming to the water defrost deposit, we had to assume that something was interfering with the water flow to that deposit. Not the solenoid valve controlling the water flow to both tanks. The tank for the ice making was full.

Water supply was checked. Water to the condenser had no problem, so the problem was to the water supply for defrosting.

By reviewing the supply line, we reached the conclusion that the filter in the water supply line for defrosts was partially clogged.

Solution

The system was de-energized. So the water supply to the machine was closed, the filter on the defrosted line water supply was removed from the line, and it was appropriately cleaned.

Finally, the entire system operation was verified.

===

EXPERIENCE 109-B

Analysis/Procedure

When I arrived at the school, I inquired what type of electric motor was used here. I found that it was a Permanent Split Capacitor (PSC) motor type. So, all I had to do was to get into the running capacitor and change the electrical connection.

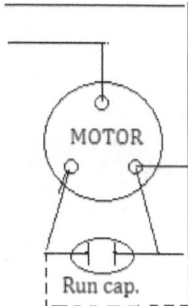

The electrical diagram at right shows the installation as it was (- - - - -) dotted line. With this electrical connection, the motor was running as a ventilator.

The electrical connection was changed: The dotted line was removed and the motor ran in the opposite direction. Extractor.

I energized the motor, and it worked as planned. All I had to do was return to the store, buy another device, then find a way to install them.

==

EXPERIENCE 110-B

I do not know what his helper's specialty or occupation was, but a final word:
"Stick with what you know."

or "If you don't know it, don't do it."

Diagnosis

I went to the place and found that they "forgot" to remove the rubber plugs at the end of every piece of rigid tubing (lengths of 10 feet each).

All the soldering connections had to be remade.

Sometime later he found a job as a design engineer in a Long Island R & A/C company.

===

EXPERIENCE 111-B

<u>Diagnosis and Procedure</u>

In order to get the exact compressor's replacement, we should take:

1. Model and serial numbers
2. Voltage values
3. Frequency (hertz)
4. Number of phases (1 and 3)
5. Size of the connecting lines and other units' characteristics
6. Equipment's application

With this information, I called them. The person with whom I spoke did not even have an idea which or what was represented by each one of the numerals (number 1 above) in the compressor's identification plate, and after a lot of struggling, I ended up adapting the compressors that they sent. The compressor's connections had different sizes.

Another important point of this training idea was that the owner's son wanted to impose his "authority," forcing the workers to make the courses and get graduated so they would be able to ascend in their categories.

The result: the workers spoke with the union representative who said that that was against the union's statutes, which that it could not be done. At the end of the entire project, it was a complete failure.

It does not matter how much we want to do something well; we do not always get whatever we desire to obtain.

When using a voltmeter to test an "open" control in an energized electric circuit, the meter must read the voltage value in that circuit.

If the electric meter used is an ohmmeter (circuit de-energized), the meter should show something other than a "continuity" reading.

Somehow, the size of the connecting lines had to be provided to the compressor's salespeople.

===

EXPERIENCE 112-B

Comment

Many times in our lives, we meet people that really make us think. And unwillingly, we learn from them.

When in college, one of my classes was suspended. Therefore, I had two free hours until the next class. I decided to pay a visit to my former student's business. It was near the college (Voorhees in Manhattan). I went to the place; I parked the automobile and walked to his shop.

Outside, a woman was cleaning a refrigerator. I approached it (behind her). I asked for the man, and she, without even turning her head around, said to me, "Mr. Jimenez." I was surprised, and I asked her, "How did you know who was I"? She answered me, "That voice, I hear it at all hours even in my dreams." She greeted me very warmly.

===

EXPERIENCE 113-B

I taught for the Refrigeration Service Engineers' Society (RSES) New York chapter, for a period of 16 years (1985–2001). During that time, I instructed the RSES members in La Guardia Community College (continuing education program) in air conditioning, refrigeration, heating, etc. (R & A/C Term's Nos. I, II, and III), Heat pumps (Term I) Electric Controls (Term I and II) areas and EPA refrigerants handling certification.

===

EXPERIENCE 114-B

Sometime later (we do not know how long), the discharge operating pressure of the refrigerant "broke*" and went through the pipe's 90° elbow; it was so the impact of the refrigerant vapor leaving, that made some damage to the wall

Technical Data

* R-22, with a normal discharge operating pressure of about 210-245 psig., has at the compressor's outlet a temperature that could be between 122 and 167°F. (sometimes even higher). The refrigerant vapor velocity between these two temperatures may be within 619.5 and 641.5 ft. / sec.

What do you think?

==

EXPERIENCE 115-B

While he was attending the school, I remembered that he always came to the class on time, and at the end of it, at about 11:00 p.m., he had to go to work at Lower Manhattan, cleaning floors in those buildings. Nor did he want a loan to pay for tuition; he paid it on a weekly basis. During the class period, he did not allow from the other participants any waste of time.

Once, I had a toothache that hurt a lot. That night after class, he took me to a dentist friend of his who extracted the tooth. I spent all night bleeding. Well, at least Santos (my landlord) and his mother-in-law were worried, and they provided some homemade remedies to me.

==

EXPERIENCE 116-B

<u>Analysis - Solution</u>

I verified the pump's electric motor and found out that water had gotten it into it; as a result, it was burned out. I took the motor pump's complete information and went to purchase a new one.

When I came back with the new water pump replacement, the place was almost completely dried out.

I proceeded to disconnect the old water pump from the piping, and found no water in the water pump's discharge line; for me, this problem happened because the mechanic did not "bleed" (purge) the pump conducts to remove the air from them. In other words, the water pump never moved any water.

And so, the place had gotten flooded. I installed the new pump; and it bled its conducts. The entire pump's operation was verified.

===

EXPERIENCE 117-B

<u>Analysis</u>

As explained at the beginning of this experience, it was springtime; the temperature in this place was higher than the outside temperature. It seemed to me as well that the place was kind of dry.

While we observed the system working, I did not know why, but a student asked me if the ambient conditions there might have something to do with the equipment operation.

I asked him what he meant by that question. He explained to me that some time ago, he was checking the operation of another A/C equipment and that he had experienced the same problem.

He thought about how in a regular operation, the A/C systems produce water and that water goes to the condenser and is used in the system's operation to help in removing the heat from the refrigerant, It occurred to him to add some water to the condenser's side pan.

I agreed with George thinking, it made a lot of sense to me. Then due to mere curiosity, we added some water—and guess what? The water helped, and almost immediately, the working conditions changed.

Applying data on tables numbers 1 and 2 next page.

1. Let's assume that water from the evaporator going to the condenser's pan (40°F.) has a heat content of 9.65 Btu/lb (table 1 first row).

2. The refrigerant vapor inside the condenser (+/-160°F) and has a heat content of 112.247 Btu/lb.

3. Temperature difference = 160 - 40 = 120°F.

4. Air passing through the condenser at 80°F (37 lb) (2.7 cu ft/m/ton)

5. Air difference in moisture content (80°°F = 155.8 - 40°F = 36.4 grains/lb)

Heat removed by water = 36.4 grains x 3.6 lb R-22 = 131 grains/min

(1 pound of water is equal to 7,000 grains) (7,000 gr/lb /16 oz/lb = 437 gr./oz)

Heat removed by water = 9.65 Btu/lb x 3.5 lb/ton = 33.78 Btu/ton

----

Heat content vapor condenser (+/-160°F) = 112.247 Btu/lb.
minus heat content water condense pan (40°F) = <u>9.65</u> Btu/lb.
                                                 102.597 Btu/lb.

==

Dry Air

At Atmospheric Pressure of 29.921 in. hg. Abs.

°F	Specific Volume cu ft lb	Density lb cu ft	Heat content (Enthalpy) Btu lb
40	12.59	0.0794	9.65
41	12.62	0.0792	9.89
42	12,64	0.0791	10.14
80	13.60	0.0735	19.32
85	13.73	0.0728	20.53
90	13.86	0.0721	21.74

Table 1

Mixtures of Dry Air & Saturated Water Vapor

One pound of dry air saturated with moisture, at a total pressure of 29.921 in. Hg. (Atmospheric Pressure)

°F	Specific Volume cu. ft./lb.	Density Lbs. /cu. ft.	Moisture content (per lb. of dry air) Pounds	Grains	Heat content (Enthalpy) Btu's./lb. of air & moisture Sensible	Latent	Total
40	12.70	0.0787	0.005202	36.4	9.65	5.56	15.23
42	12.76	0.0764	0.005625	39.4	10.14	6.01	16.14
45	22.85	0.0778	0.006322	44.2	10.86	6.73	17.59
80	14.09	0.0710	0.02226	155.8	19.32	23.31	42.64

Table 2

496

Formulas

$$\underline{16,000} \text{ Btu/ton /hr.}$$

6. Amount of air calculation: $Q = 60 \text{ min/hr.} = 267 \text{ Btu/ton /min.}$

a. $Q = W (Dt - \Delta t) \text{ Sh.}$

$$W = \frac{Q}{(\Delta t.) \text{ Sh}} \quad \frac{267}{30 \times 0.24} = 37 \text{ lb/m/ton} \times 0.0735 \text{ lb/cu ft} = 2.7 \text{ cu ft/min/ton}$$

b. R-22 Enthalpy at 80°F.

Vapor 111.239

Liquid $\underline{33.342}$

77.90 Btu/lb.

x $\underline{0.73}$ Flash gas

57.0 Btu/lb = 77.90 - 21 = 57.0 Btu/lb.

$$\frac{200 \text{ Btu/min/ton}}{57 \text{ Btu/lb}} = 3.5 \text{ lb/min/ton}$$

===

EXPERIENCE 118-B

The following Monday afternoon, I called the telephone number that the instructor gave me; he let me know that he was really surprised. He said that even without me attending the classes, he had never had an examination result as satisfactory as mine. He told me that the grade had been 98%.

Only out of curiosity did I ask him if he could tell me the questions that I had failed.

He enumerated them to me, and they were exactly the two questions in which I used more time and that I could not answer.

==

EXPERIENCE 119-B

I went to the ceiling, where he told me the equipment was located. I saw a colored spaghetti plate-like installation, and I did not find any heating printings. By looking at the equipment covers, it seemed that somebody had removed them. So I went down and spoke to the owner about the heating system's printings; he did not know anything about them. Then, I noticed that in the bar location, the temperature was right. I asked where the equipment supplying heat to that zone was located.

They gave me the idea that it was right on the ceiling above. I climbed up to the ceiling, and I saw right there a heating system similar to the one with the problem. I looked around, and I was able to get the printing plans for that system. They were attached to the equipment's cover. Because if I removed the cover, I had to stop the system, I exchanged the covers. I made a note indicating where the printings could be located and attached to the system's cover.

I got back to the other ceiling location, and by using the printing plans I was able to put the troubled equipment back in operation.

Time of the repair: about one and a half hours.

I verified the entire operation of the system, and after everything worked well, I left the place.

===

EXPERIENCE 120-B

Comment

I think this experience is self-explanatory.

The important thing is to apply yourself to accomplish what you want.

Everything involves some kind of sacrifice and effort, but in the end, we can harvest the fruits of it.

==

EXPERIENCE 121-B

<u>Procedure</u> (printing plans)

Note: Printing plans and electrical installations (in any electrical circuit) can be followed in two different ways:

1. Forth. Beginning at the power supply, following through the controls; toward the power-consuming devices, or
2. Back. From the power consuming devices through the controls to the power supply inlet.

I started following the three live lines one at a time toward the main high-voltage components: fuses, contactors/starters (unit (compressor-electric motor), evaporator/condenser fan motors, crankcase heaters, etc.). After about fifteen minutes and following all three lines one by one, I found everything normal in that circuit.

Then, I started to verify the 24-volt circuit control lines. As cited above, I stared at one end of the power supply (transformer), followed the line through the controls (no loads) until I got into the voltage load (starters-contactor's coils, etc.). So far, in that line, there was no problem.

Now, the other 24-volt supply control line. I started again, this time on the timer control. Here, I found there was no connection to the timer's motor or to its distribution control contacts.

Since this device controls the entire equipment operation, due to the missing line, nothing was working. Due to this missing wire connection, there was no power supply on the unit, the fan's contactor's coils, crankcase resistance (limit switch), thermostat, etc.

I let him know my findings. I remember him saying, "Two heads think better than one."

Note: *When working with electric diagrams and/or in the equipment's power or voltage supply lines, it is very important NOT to go through the loads; this way we avoid any mistakes. That's what happened here.*

On the other hand, electrical circuit components can be surely verified by using electrical meters:

    System de-energized = Ohmmeter "Continuity-openness-resistances-etc."
    System energized   = Voltmeter "Power reading-no reading."

===

EXPERIENCE 122-B

Solutions

In some A/C equipment, the manufacturer has already taken care of this problem by installing a tube, as shown in the figure below. But in the field, we may have to do it by ourselves.

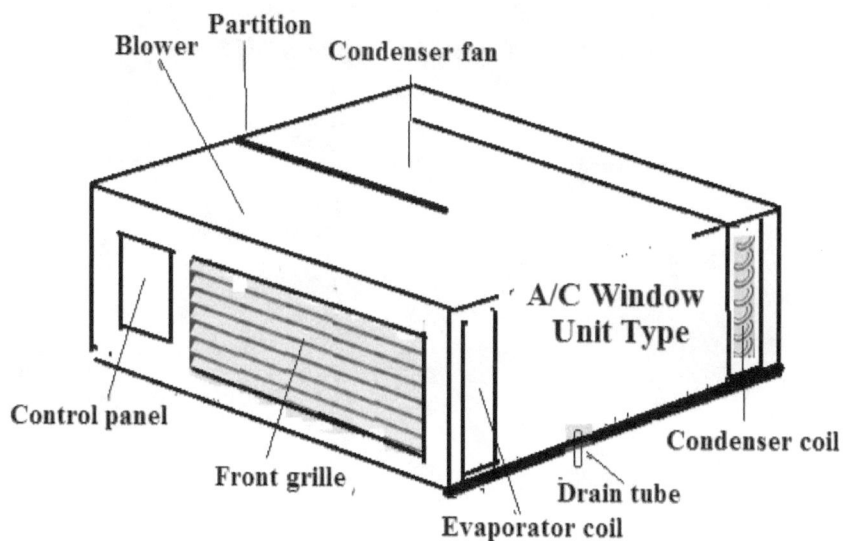

To minimize the amount of water dripping from the equipment, without affecting the system's operation, the following procedure can be exercised.

a) Drill a hole to the condenser's pan (see figure above)

Introduce a piece of copper tubing (3/8 inch) to a point where the upper end of the tube gets to about 3/8 inch below the pan's edge, and welded. Use a rubber or plastic hose, attached to the lower part of the tubing, to take the excess of water from the pan to the ground.

To accomplish the "use" of this water, the manufacturers often provide their systems with two devices or

Mechanisms

1) Splashier ring (hoop). It is attached to the tip of the condenser's fan blades. When the blades rotate, it sends the water from the pan to the condenser's tube fins.

2) Shaft equipped with a hanging belt. When rotating, it picks up the water from the pan.

Note: *See Experience 117 ("A/C problem") can be used to solve water amount calculations.*

===

EXPERIENCE 123-B

<u>Procedure</u>

I started the job by verifying each electrical component resistance-ground conditions by means of an ohmmeter. One by one (motor compressor, evaporator's motor, thermostat, timer, crankcase's resistance, controls, etc.), they were giving normal readings of resistance/closed or opening, and no one showed "ground" reading. I verify their existing electrical connections as well.

I took a measurement of the wire's sizes needed in the high voltage (230 volts) (AWG 10) as in the control circuit (24 volts) (AWG 14); then I calculate the amount of wire required of each size. I gave the list of what it was needed to the person in charge of the place.

About one hour later, the materials arrived, and I started the job.

It took me, about four and a half hours to finish the wiring.

They provided me with a very frugal lunch and a coffee.

At about three o'clock, I installed the gauges and verified the equipment's entire operation.

In the end, they signed my papers and completion of the job and gave me a hundred dollars' tip.

===

EXPERIENCE 124-B

Explanation

I think this experience is self-explained. It was one of those occasions in which we had the opportunity of learning something that does not happen every day.

It is clear, that sometimes people judge actions and situations without knowing the complete facts. But we were exposed to the real thing, and the explanation came from someone we knew very well who explained the truth.

===

EXPERIENCE 125-B

No, I did not go to Detroit.

Explanation

One day, I had the car parked in front of the school at 117th Street and 2nd. Avenue. I had opened the hood and was trying to find by myself the cause of the stalling problem.

The truth? I didn't have the slightest idea what to look for. It could be a loose electrical connection? I didn't know.

While I was there trying to solve the "mystery" by myself, a strange Spanish-speaking man approached me from behind, and he asked, "Are you experiencing stalling problems with your car?"

Solution

I turned my head to look at him, and said, "Yes, that's exactly the problem." He told me, that he was an auto mechanic and he knew what the problem was; then, he asked me for a two-foot piece of copper wire #12 AWG, a pair of pliers, and a screwdriver. I gave him what he asked for, and he proceeded to change a wired connection in the auto-electric installation.

It took him about five minutes. When he finished, he gave me back the tools and the piece of wire that he had removed and explained to me that that piece of wire was made of aluminum and at certain temperatures, it experienced some change and acted as a resistance, blocking the current flow, producing the problem.

I was so thankful for this man. He just said welcome and went away.

We never know when least expected, someone comes to help. Indeed, as far as I know, aluminum wire is not that popular in electrical installations. Sometime ago, I heard or saw an adverse commentary about this metal used in R & A/C electrical circuits, which is also used in many parts of refrigeration and air conditioning equipment. But since I do not remember exactly what it was, I prefer not to comment.

Hold it. I remember a sign on an electrical circuit: Do not use aluminum wire! That I saw on an A/C equipment. This metal may be good for certain applications, but I prefer to work with it as few times as possible.

==

EXPERIENCE 126-B

<u>Analysis</u>

After getting familiar with the entire refrigeration system's components and before doing anything else, I conducted the following procedures:

1. Since this system uses a water-cooled condenser, I thought, *Let's check the condenser first.* Through a drain valve, some water from the condenser's outlet was taken into a small bucket; an electronic leak detector sensor tip was passed over the water surface but no sign of refrigerant leak was found there.

2. By using the electronic leak detector, the high-pressure control and the water-regulating valve mechanisms (bellows, etc.) were verified. No leaks were detected.

3. Make sure the suction service valve was positioned midway, to allow the refrigerant pressure signal to the low-pressure control.

4. Using the same tool and procedure as in number 3, the low-pressure mechanism (bellows, etc.) was verified. As soon as the leak detector's sensor tip approached the named parts, a refrigerant leak appeared.

I assumed that the leak tests on these parts of the system were neglected. Somehow, they forget that these bellows are also part of the refrigeration system.

Solution

The low-pressure control was changed. Then the entire system operation was verified, refrigerant was added, and pressure controls; were adjusted.

===

EXPERIENCE 127-B

Diagnosis.

When installing the manifold gauge to the high-pressure side (150 psig.), I noticed the pressure, compared with the room temperature, was low. This, under this circumstance, may be an indication of a lack of refrigerant.

When observing the upper part of the compressor's sides, oil was noticed on its surface.

When applying the refrigerant's leak detector in these points, the device reacted, showing an evident refrigerant leak around the compressor's cover and valves plate.

Up to this point: a deficient compressor's packing.

Making use of the appropriate wrenches, the screws securing the cover and the valve's plate to the compressor's body were tightened. But the refrigerant leak did not stop.

The possible explanation for this refrigerant leak should be the high temperatures of the place, which can create, with R-502, very high discharge operating pressures (to up to approximately 290 psig.).

Since oil was lost through the leaking gaskets, the rest of oil from the crankcase was drained out.

By closing front-seated the "king" valve located at the receiver's outlet, the system was submitted to a pumped-down process until it got a gauge pressure reading of 0 psig.

Solution

Change the entire compressor's gaskets set.

1. *Normally, when the original packing is installed, the pressure exerted by the fastening screws cause the packing to be compressed between the two mechanical parts, to make sure the seal between them. When the packing is removed, or released (as in this case), it loses its sealing ability and retightening no longer works. So the best solution is to change it.*

2. *To keep the equipment's operating pressures within a normal level would be recommended, as well as the installation of a water-cooled condenser in a series circuit with the existing air-cooled condenser.*

3. *Since the gaskets' failure was caused, by excessive pressure, in order to prevent this problem from happening again, it should also recommend the installation of a suction pressure regulator. But points 2 and 3 were up to the client.*

I talked to the client, explained the entire problem, and recommended the installation of the water-cooled condenser, the new WRV, as well as the installation of the SPR, the charge of new oil, and the change of the compressor's gaskets set. The customer considered my explanation, agreed with it, and gave the OK to the repair.

The manufacturer's information from the compressor's ID plate was taken: model number, serial number, electrical characteristics, compressor's capacity, etc.

A new 1/2-ton water-cooled "tube within a tube" type condenser and a new water-regulating valve, the suction pressure regulator, as well as the gaskets set were obtained.

Carefully, the old gaskets were removed. The new gasket set was wet with fresh oil and installed. The new water condenser was connected in a series circuit with the existing air-cooled one, and the new water valve was connected. The SPR was connected to the system's suction line at about one foot from the compressor. The existing filter-drier on the liquid line was replaced with a piece of tubing and attached with flare nuts.

After this part of the job was done, the equipment was pressurized with R-502 and verified for refrigerant leaks.

The system was submitted to a forty-five-minute vacuum procedure. The proper amount of oil was added by using the vacuum in the compressor. Some R-502 was added until reaching 5 psig while the refrigerant was leaking, and the new filter drier was installed and checked for leaks.

Charging the refrigerant, adjusting the water-regulating valve and the suction pressure regulator and verifying the entire system operation ended the job.

==

EXPERIENCE 128-B

Calculations (R-134a)

According to the characteristics of this refrigerant, the operating pressures should be the following:

a) *Discharge pressure:*

Condensing temperature = Ambient temperature = +/-80°F.
+ Temp. diff. (Δt - Dt) = 25–35°F.
+/-105–115°F.

Discharge pressure = 135 to 158 psig.

b) *Suction pressure:*

(B. P.) Boiling temperature = Desired temperature = 15°F.
- Temp. diff. (Δt - Dt) = 15°F.
0°F.

Suction pressure = Approximately a fluctuating 9 psig.

c) EPR adjustment: For this calculation, look at Experience # 101.

Diagnosis/Procedures

(1) The EPR was adjusted to 10 psig. (7°F.). Too low.

(3) The high-pressure control's settings were found OK, although the low-pressure control operative condition was found defective.

<u>Solutions</u>.

The EPR was adjusted at approximately 33 psig (38°F).

The low-pressure control was changed and adjusted for the application.

The entire system's operation was verified.

===

EXPERIENCE 129-B

<u>Analysis</u>

Since the required information was for only one ice container, the rest of the system information was up to him.

I limited myself to showing him the calculation on the amount of heat required to be removed in order to cool-freeze the specific amount of water:

(1) (3) Qs. Sensible heats values above and below the freezing points
(2)   Ql. Latent heat values.

According to the results, the assumed system's capacity.

		°F		°C
		60		16.15
(1) Qs. = W (Δt.- Dt.) Sh. (water = 1 Btu's/lb)			(1)	
1,498 (60 – 32) 1				
= 1,498 x 28 x 1	= 41,933 Btu's.	32		32
(2) Ql. = W (LHF) (ice = 144 Btu's/lb)			(2)	
1,498 x 144	= 21,571 Btu's.	32		32
(3) Qs. = W (Δt.- Dt.) Sh. ice = 0.504 But's/lb)			(3)	
1,498 (32 – 18) 0.504		18		-7.7

Qt. = (1) Qs. = 41,933
    (2) Ql. = 21,571
    (3) Qs. = 10,570
       Qt. = 74,074 Btu/8 hr = $\frac{9,260}{12,000 \text{ Btu/ton/hr.}}$ Btu/hr.

                       = Approx. ¾ ton.

He didn't ask for anything else. But if we wanted to know about the refrigerant to be used and the equipment normal operating pressures, experiences 69 and 101 can be consulted.

===

EXPERIENCE 130-B

Analysis

In an agreement with the information provided by the client and the equipment's appearance, the compressor should be tested before taking the decision of changing or repairing it.

Comment.

For me "Not old enough," the compressor must be put under other tests before deciding if it must be replaced or repaired.

The obtained operating pressures indicate compression deficiency; this, together with the "noise" can be indications of something loose or broken within the device.

The temperatures of the liquid and suction lines were also an indication of deficiency in the circulation of the refrigerant through the system.

As the pressure gauges are already installed to the system, the next step should be to verify the pumping conditions or the compressor's displacement.

The procedure "testing of compressor's efficiency" is advisable to be performed here to verify its operative mechanical condition.

If the compressor fails to pass this test, one of the following options can be considered:

1. Change of the compressor
2. Intent repairing the compressor (semi hermetic or open types)

Note: The repair step in number 2 involves isolating the compressor, disassembling it, and verifying its internal mechanical components for change:

Suction and/or discharge valves or the valves plate set, and in addition, the change of the entire packing gasket set.

Regardless of the selected option, we must try NOT to expose the compressor/system to environmental conditions.

Before deciding to either change or repair, I chose to isolate the compressor of the system by closing front-seated the compressor's two service packing gland type valves; once the compressor pressure was released, the two cylinder's covers were removed as to observe the internal valves' condition.

In both sides of the V-type compressor broken valves were found, its pieces moving freely in the space above the cylinders. This finding answered the question of the compressor's noises.

After explaining and commenting with the customer about the founded condition, option 2 was adopted. The pertinent information of the equipment was collected, and the appropriate set of gaskets and valve's plates were ordered from the spare parts distributor.

The EPR at the higher temperature evaporator's outlet setting was verified.

Solution

Once the corresponding spare parts were obtained, following the appropriate procedure, the change was performed.

The compressor's oil charge was removed and the required amount verified.

The repair was completed; the compressor was pressurized and verified for leaks. It was put under a vacuum process for a period of one hour. Making use of the compressor's vacuum, the charge of oil was added, the service valves were set in their normal working position, and the system started to work.

The pressures, temperatures, and amperages of operation were verified.

For EPR adjustment and pressure and temperature calculations, see Experiences 101 and 128.

===

FORMULAE & OTHER CALCULATIONS

"R-A/C OPERATING PRESSURES"

Here, are contained some formulae and calculations used in RA/C & E to estimate operations and procedures carried out in these trades fields.
--}

"SUCTION (Back - Low) PRESSURE"
Key = HEAT FLOW
==|

The (L.P.) "Low – Back or Suction Pressure" of a Refrigeration or Air conditioning system, is defined as:

"The operating pressure of a system, measured in the suction line at the compressor's inlet".

The purpose of this pressure is:

"To create the necessary temperature differential (TD - Δt.) (Refrigerant – Space/material) in order to accomplish the appropriate cooling effect"

This operating pressure is mainly affected, by the "HEAT LOAD" value required in the space or material to be cooled, although

This pressure calculation value, depends on the following factors as well:

1. Type of refrigerant used in the system:

   Refrigerants differ in their chemical structure, for this reason, their operating pressure's values, are different.

2. Characteristics of space – material to be cooled {application (*)}, requiring different conditions. I. e: temperature, moisture content, others.

    (*) Application = High, Medium and Low temperatures.

3. Temperature difference (Δt. -DT), required accordingly with application.

    Space/material → Refrigerant:

    In *all applications,* this temperature difference it is calculated, considering the cooled material requirements.

    a. A/C for human comfort about 72 °F. (Δt.) Differential = 30 to 35 °F.
        In other A/C works this variable depends on the characteristics of the application.
    b. Commercial refrigeration, the type of product as well as its storage requirements.
        If this is not taken in consideration, adverse effects as: Freezing, de-hydration, weight loss, appearance, others, may result in damaging or spoiling the storage product.
    c. Other equipment mechanical operating conditions.

To verify the equipment normal operating pressure, two factors should be remembered:

1. We most know, what should be the appropriate operating pressure for that specific system.
2. What the pressure reading is, at the time of the service call.

After comparing the two conditions, a more accurate diagnosis about its status can be reached.

═══|

C A L C U L A T I O N S

SUCTION OPERATING PRESSURES

A/C Systems

Evaporator Pressure (Boiling Point)

Desired Temperature = 72 °F. (22.2 °C.)
(minus) (-) Differential = 35 °F. or 30 °F. (1.7 or -1.1 °C.)

Boiling Point (B.P.) = 37 °F. to 42 °F. (2.8 to 5.6 °C.)

Changed to Pressure (R-22) = (37 °F.) = approx. 62 Psig. (Min. Norm)
(42 °F.) = approx. 72 Psig. (Max. Norm)

═══|

Commercial refrigeration

B. P. =   Desired Temperature
        (minus) - Differential   (Δt.)   (According to application) **
        (Refrigerant)   Boiling Point (B. P.)

**   Desired Temperature   = _____ °F. _____ (°C.)
        (minus) - Differential (Δt.)   = _____ °F. _____ (°C.)
        (equal)   Boiling Point (B.P.)   = _____ °F. _____ (°C.)

Differentials: Hi. Temperature = (35 - 30 °F.) (16.7 - 19.4 °C.) (A/C.)
        Med. Temperature = (10 °F.)   (5.6 °C.) (Vegetables & Fruits)
        Med. Temperature = (15 °F.)   (8.3 °C.) (Canned-Contained Products)
        Low. Temperature = (+-15 °F.)   (8.3 °C.)

---|

(Δt. - Dt.) = Delta or temperature differential.
(B.P.)   = Boiling point.

MINIMUM-MAXIMUM NORMAL PRESSURE READINGS

| (Min.) as close to the "FP" (Freezing Point) of water [32 °F.
(0 °C.) as possible.

Boiling Point (B.P.)   > (Max.) 48 °F. (8.9 °C.) @ max. Condensing
Temperature of (Med. temp.) | 130 °F. (54.4 °C.)

|

--|

Sample 1 Vegetables

Desired Temperature = 48 °F. (8.9 °C.)
(minus) (-) Differential = 10 °F. (-12.2 °C.)

Boiling Point (B.P.) = 38 °F. (3.3 °C.)

Using R-134 = (38 °F.) = approx. 33 Psig.

--|

Sample 2 Beer cooler.

Desired Temperature = 40 °F. (4.4 °C.)
(minus) (-) Differential = 15 °F. (-9.4 °C.)

Boiling Point (B.P.) = 25 °F. (-3.8 °C.)

Using R-134 = (25 °F.) = approx. 22.0 Psig.

--|

Sample 3 Ice maker.

Desired Temperature = 32 °F. (0 °C.)
(minus) (-) Differential = 15 °F. (-9.4 °C.)

Boiling Point (B.P.) = 17 °F. (-8.33 °C.)

Using R-134 = (17 °F.) = approx. 3.0 Psig.

--|

Note. For other type of refrigerants, the same temperature differentials
can be used.

==|

"DISCHARGE (Head - High) PRESSURE"

Key = HEAT FLOW

==|

The (H.P.) "High – Head or Discharge Pressure" of a Refrigeration or Air Conditioning system, is defined as:

"The operating pressure of a R-A/C system, measured in the discharge line at the compressor's outlet".

The purpose of this pressure is:

"To create the necessary temperature differential (TD - Δt.) (Refrigerant – Cooling medium) in order to accomplish the refrigerant's condensation"

The main mechanical component of the system is the Condenser.

This operating pressure is affected, by diverse factors:

01. Type of refrigerant used in the system.
02. System characteristics application
03. General characteristics of materials used in the condenser's design.
04. Cooling mediums substances characteristics:

   a) Temperature.   b) Source.   c) Amounts.
   d) Geographic location.   e) Others.

05. Refrigerant - Cooling medium temperature differential (Δt. -DT),
06. Weather conditions
07. Type of compressor. Open, Hermetic or semi-hermetic.
08. Other

The condenser's manufacturer designs it, taking into account the points cited above.

Heat eliminated in the condenser:

a. Super heat   b. Condensation   c. Sub-cooling

As with the suction pressure, to verify the equipment normal operating discharge pressure, two factors should be remembered:

1. The appropriate normal operating pressure for the system should be known.
2. The actual pressure reading at the time of the existing problem.

Then through the comparison of these pressure readings, a more precise diagnosis can be obtained.

===|

In the R & A/C fields, must of this factors are, or can be affected/controlled, up to certain point, others (*), not.

(1) <u>System's capacity</u>. Usually, this calculation is performed, by using a one-ton system's capacity, and within an approximate 10% variation. Meaning 20 Btu's /ton/ min. or 1,200 Btu's/Ton/hr. or 1 Ton = 12,000 +- 1,200 Btu's/Ton/min.

Meaning 200 – 20 = 180 or 200 + 20 = 220 Btu's/min/ton or
1 ton = 12,000 - 1,200 = 10,800 or 12,000 + 1,200 = 13,200 Btu's/hr./ton.

(2) (*) Type of metal (Copper, Bronze, Iron, Aluminum, Etc.) and other constituents used.
I.e.. Finns, vanes, extensions, Etc. Each metallic material has different specific heat values.

(3) Different sensible heat, specific heat, specific volume, density's values, Etc. of the cooling medium substance used.

I.e. Specific heat (Btu's/°F. /min.) = Water = 1   Air = 0.24   Glycol = Approx. 0.84

(4) Due to difference of the various facts cited at number 3.
(5) This, affected by temperature differences, densities and etc.
(6) Climate, geographical locations, time of day, etc.
(7) And other unexpected factors.

---|

Some Specific volume, Density and Heat content
values in air at various temperatures

At Atmospheric Pressure of 29.921 in. hg. Abs.

Temp.	Specific Volume	Density	Heat content (Enthalpy)	Specific heat
°F.	cu. ft. lb.	lbs. cu. ft.	Btu lb.	Btu /lb. / °F.
70	13.35	0.0745	16.82	This value
75	13,48	0.0738	18.02	fluctuates
				around the
80	13.60	0.0735	19.22	number
85	13.73	0.0728	20.53	*0.2630*
				@ At. press.
90	13.86	0.0721	21.74	14.696 psi

Calculations.

Airflow – Water flow – Etc. Some of the following formulae can be used for these calculations.

CFM = W x Sp. Volume

Where:

CFM = Cubic Feet per Minute.
W = Weight. Pounds.
Sp. Volume = Specific Volume. Cubic feet/lb. (See Table below)
---|

Qs. = W (Δt. - Dt.) Sh.

Where:

Qs. = Sensible heat. Btu's/ton/min.

527

W = Weight. Pounds.
Δt. - Dt. = Temperature difference. Deegres °F.
Sh. = Specific heat. Btu's/ lb./ °F.

Ql. = W x LHV.

Where:

Ql. = Latent heat.
W = Weight. Pounds.
LHV = Latent heat of vaporization.

Qs. = A (Δt.) "U" Factor. $\qquad A = \dfrac{Q}{(\Delta t)\ \text{"U" Factor}}$

$$W = \frac{Q}{(\Delta t. - Dt.)\ Sh.} = \frac{200\ \text{Btu's}/\text{T}/\text{min.}}{20 \times 1} = +\!-\ \frac{10\ \text{lbs.}/\text{T}/\text{min.}}{7.8\ \text{lbs.}/\text{gal}} = \text{Approx. 1.2 Gall. /T. /min.}$$

8.34 lbs. = 1 gallon of water.

$$7.48\ \text{galls./cu. ft. (water)} = \frac{62.4\ \text{galls./cu. ft.}}{8.34\ \text{lbs./gall.}}$$

GPM. 1. (Municipal)   +/- 20 °F.   2. (Cooling tower) Δt. 10 – 12 °F.

$$W = \frac{Q}{(\Delta t.)\ Sh.} \quad \frac{267\ \text{btu's / ton /min.}}{(20\ \text{F.})\ 1 = 13.33\ \text{lbs. /ton /min.}}$$

About the Author

Jose C. Jimenez was born in Bogota, Colombia, in South America.

He began his HVAC & R and electrical working and teaching career in his country in 1958 while serving in the Colombian navy and continued it in the United States. Here he has worked as a service technician. He was instructor-director in several trade schools (National Skills Center, Technical Trade School Inc., GM Tech Inc., NY La Guardia Community College, HVAC Tech Inc. School, Refrigeration Institute, etc.) He has written varied HVACR and electrical textbooks, which had been used in several of these schools, as well as preparation for EPA certification, and NYC Fire Department RMO licensing. He has been affiliated with organizations such as ASHRAE, RSES, and NATE. He has an Associate Degree in Applied Sciences in Environmental Control Technology.